Consulting Editor: E. A. Reeves

ELECTRICAL VARIABLE SPEED DRIVES

K. G. Bush

Blackwell
Science

© 1995 by
Blackwell Science Ltd
Editorial Offices:
Osney Mead, Oxford OX2 0EL
25 John Street, London WC1N 2BL
23 Ainslie Place, Edinburgh EH3 6AJ
238 Main Street, Cambridge
  Massachusetts 02142, USA
54 University Street, Carlton
  Victoria 3053, Australia

Other Editorial Offices:
Arnette Blackwell SA
1, rue de Lille, 75007 Paris
France

Blackwell Wissenschafts-Verlag GmbH
Kurfürstendamm 57
10707 Berlin, Germany

Blackwell MZV
Feldgasse 13, A-1238 Wien
Austria

First published 1995

Set by Setrite Typesetters, Hong Kong
Printed and bound in Great Britain by
  Hartnolls Ltd., Bodmin Cornwall

DISTRIBUTORS

Marston Book Services Ltd
PO Box 87
Oxford OX2 0DT
(*Orders*: Tel: 01865 791155
         Fax: 01865 791927
         Telex: 837515)

USA
Blackwell Science, Inc.
238 Main Street
Cambridge, MA 02142
(*Orders*: Tel: 800 215 1000
         : 617 876-7000
         Fax: 617 492-5263)

Canada
Oxford University Press
70 Wynford Drive
Don Mills
Ontario M3C 1J9
(*Orders*: Tel: 416 441 2941)

Australia
Blackwell Science Pty Ltd
54 University Street
Carlton, Victoria 3053
(*Orders*: Tel: 03 347-5552)

A catalogue record for this title
is available from the British Library

ISBN 0 632 02226 4

Library of Congress
Cataloging-in-Publication Data

Bush, K. G.
    Electrical variable speed drives/
  K. G. Bush: consulting editor,
  E. A. Reeves.
      p.   cm.
    Includes index.
    ISBN 0-632-02226-4
    1. Electric driving, Variable
  speed. I. Reeves, E. A.
  II. Title.
  TK4058.B84   1995
  621.46 − dc20   94-38727
                                      CIP

Contents

Preface   v

1 Introduction   **1**

2 Electric motors   **5**
   D.C. motors   5
   A.C. motors   14

3 Open and closed loop control   **25**
   Feedback   26
   Open loop control   27
   Closed loop control   27
   Examples of control systems   28
   Characteristics of control systems   31
   Performance specification   34
   The application of closed and open loop control   35

4 Power semiconductors   **37**
   Diodes   38
   Transistors   40
   Thyristors   42
   Gate turn-off devices (GTOs)   44
   Protection of semiconductors and loads   45
   Semiconductors in inverters   47
   Semiconductors in converters   49

5 Variable speed using a.c. motors   **53**
   Induction motors   53
   Pole changing   56
   Series resistance control   56
   Specially designed a.c. motors   56
   Slip recovery   59
   Static Kramer system   60
   Variable frequency control   61

6   Variable speed using d.c. motors   **69**
Field control   69
Armature resistance control   71
Choppers and pulse width modulation   71
Ward Leonard-Ilgner control   71
Single-phase thyristor converters   76
Three-phase thyristor converters   77
Regenerative braking   80

7   Harmonics   **83**
Harmonics and drives   83
Electromagnetic compatibility (EMC)   89
EU directives   90
Standards   91
Conducted and radiated emissions   92
Installation   94
Competent bodies   94

8   Comparison of variable speed drives   **97**
A.C. drives   97
D.C. drives   98

9   Selection of variable speed drives   **101**
Load characteristics   101
Environment   118
Maintenance   119
Cost and running cost   119
Individual drives   121
Grouped or sectional drives   122

10   Trends   **126**
Modular design   126
Digital technology   129
Improved power factor   131
Future performance of drives   131
New devices   131
Drive communications   132
The future   132

Appendix: Conversion data for drive calculations   133

Glossary   134

Index   137

Preface

Many books deal with the design and construction of electrical variable speed drives. The majority of these are intended for engineers whose interests lie in the design, manufacture and supply of this equipment. This book is intended to provide a broad view of the various types of electrical drives available for normal industrial uses and give an indication of the salient features which characterize the various types. It is for those whose main concern is understanding the basic differences in the various types of drive in current use, whether employing the latest technology or not. No attempt is made to describe all the types available; only those which are generally employed in power applications are examined.

The various features of each type of drive are explained and the language and terms used throughout are intended to make the book easily read and understood by anyone who is involved in selecting, using and living with electrical variable speed drives. It will also be of interest and use to engineers, managers and students who, whilst not directly concerned with the utilization of such equipment, require a fundamental knowledge and understanding of the function and range of the types of drive likely to be encountered in industry.

After dealing with the basic question, 'Why do we need variable speed?', the book traces the development of methods by which this has been achieved over the years and up to the present day, and provides information on the types of electric motor that are utilized in variable speed drive systems. Many simple illustrations of the performance and basic connections of the drive systems described have been included in order to support the text and bring the reader to a quick appreciation of the differences between the various types of drive.

The process of the selection and specification of drives is examined, together with worked examples of the method of determining the appropriate sizing and selection of a drive system for a typical industrial application. Legislation on the subject of electromagnetic interference and the techniques used to achieve compliance with the requirements are explained. The emergence of new technology and techniques is discussed and the accompanying advantages and potential disadvantages for the user are explored.

Chapter 1
Introduction

In recent years the use of variable speed drives has become so commonplace that we tend to accept their presence on industrial processes without question; there are, indeed, very few continuous processes that are now performed without employing them. With this in mind it may seem superfluous to suggest that before considering in any detail the various types of electrical variable speed drives available we should first consider the question 'why variable speed?'. The answer may be obvious in some instances; for example, the discharge of material from a conveyor can be most simply controlled by slowing it down or speeding it up or, in extreme cases, stopping or even reversing it.

In some other cases, however, the need for variable speed is not always so self evident. The process of changing the flow of a liquid in a pipeline by means of a valve would seem to be a simple and adequate method but when large volumes are being handled or smooth flow in the pipeline is required consideration must be given to more efficient methods of achieving the objective. Similarly, the ability to change smoothly from one operating speed to another, in an industrial process, might be considered highly desirable, particularly if the alternative would be a shut down of the process with the inevitable accompanying loss of production.

Since the industrial revolution, engineers and designers have recognized the need for variable speed control and many ingenious methods have been devised. Perhaps the simplest was a pair of cone pulleys (see Fig. 1.1) with some form of belt shifting mechanism to alter the position of the belt connecting the pulleys and hence the ratio of input to output speed. This form of variable ratio drive is limited in the range of ratios that can be achieved, for if the taper on the pulley is too great the belt position tends to change with change in load. This form of simple mechanical control of speed was extensively used and a similar

1

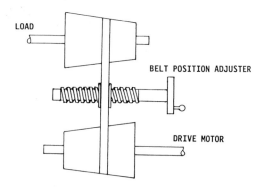

Fig. 1.1 Achieving variable speed from a constant speed source by means of cone pulleys, speed variation limited by length of pulley and slope of taper which normally does not exceed 15%.

principle was employed utilizing variable pitch pulleys (see Fig. 1.2), and the disc and ball adjustable ratio device (see Fig. 1.3). The majority of these simple mechanized systems were, however, limited in their application by their inability to transmit large powers; thus hydraulic and electrical systems became the choice for the larger powered applications.

Electrical variable speed drives were still undergoing development in the early 1900s and the majority of applications were satisfied by motor generator sets and direct current (d.c.) motors. As alternating current (a.c.) became the standard type of electrical

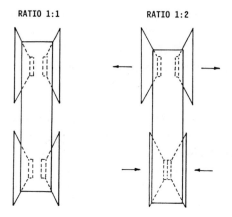

Fig. 1.2 Achieving variable speed from a constant speed source by means of variable pitch pulleys; ratios of 3:1 can be obtained in standard designs.

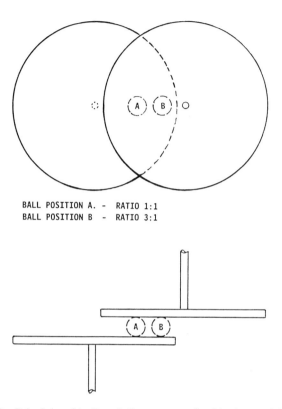

BALL POSITION A. - RATIO 1:1
BALL POSITION B - RATIO 3:1

Fig. 1.3 Principle of ball and disc system of achieving variable speed.

supply the conversion to d.c. was relatively inefficient (particularly when compared with modern methods). As three electrical machines are required – a constant speed motor driving a variable voltage generator which finally provided power to the drive motor – the combined efficiency of such sets rarely exceeded 75%.

Many mechanical, hydraulic, electromechanical and electrical variable speed drive designs have emerged and the remarkable growth of interest in this technology can be traced to the increasing cost of energy and the need to conserve resources. It is in this particular area that speed control can make a considerable contribution.

Recent developments in the design and utilization of electrical components have resulted in a significant growth in the availability of electrical speed controlled drives. Changes in this technology have been extensive, and are receiving much publicity and attention. It must, however, be remembered that there is still a place – including one for specialized applications – for the earlier technology and designs, many of which continue to perform so well that their continued application is assured.

When reviewing various drive systems it is important to bear in mind that a properly engineered drive can control torque and speed to a process, in addition to varying or even reversing the direction of rotation. This provides the user with wide ranging control over the parameters of the process to which it is applied.

It is unlikely that all fixed speed drives will be replaced by variable speed types but it is a significant fact that with the reduction in relative cost of modern variable speed drives there has been a very large growth in their application. This trend is expected to continue and, with utilization at such high levels, manufacturers will certainly give increasing attention to further improving functionality and reducing the costs.

There is no perfect drive type for all applications and the selection of a drive type or system must be made only after careful consideration of all appropriate criteria. Consideration of the load type is of primary concern and some typical applications are described in Chapter 9. Of equal importance are the basic characteristics possessed by the various types of drive. In the final analysis reference should be made to the manufacturers for data but it is appropriate that those factors which influence choice be given some consideration in general terms. In the larger power sizes the choice becomes more limited due to the electrical design problems associated with control and regulation of large electrical currents. Speed range is another consideration which determines selection as some drive forms are limited in their ability to work over very wide speed ranges. Accurate control of preset speed using an open loop control is only available in variable frequency drive systems employing synchronous motors. Other drives require closed loop feedback.

In selecting a drive for continuous operation over long periods of time reliability and efficiency are important, whereas first cost would be the main factor for equipment with a limited lifespan. In special environments choice might well be limited to equipment designs able to survive and operate successfully in harsh conditions.

Chapter 2
Electric motors

Electric motors can be divided into two groups: those for operation on a.c. supplies and those for use on d.c. supplies. All electric motors, however, share the basic principle of depending for their action on the interaction of magnetic fields.

The electricity supply systems in most parts of the industrial world are usually a.c. because of the ease of generation and transmission, and the ability to transform from one voltage to another. For this reason the majority of industrial applications for electric drives utilize a.c. supplies.

To achieve variable speed this supply is either changed to d.c. or modified by frequency changing or interposing some supply regulating device which controls speed by adjusting the current supplied to the motor and hence the motoring torque, the speed then falling or rising to that level at which the power demanded is equal to the power supplied.

D.C. motors

For the majority of variable speed drive applications the d.c. motor has been for many years the principal machine utilized due largely to its flexibility of performance. To meet the various requirements of torque and speed control a number of variations in design have evolved. These can be broadly divided into three groups, each with distinct characteristics: series, shunt, and compound wound motors.

Series

The series motor is so described because the field system or excitation is connected in series with the armature winding so

Plate 2.1 Modern d.c. motor with end-frame-mounted separately-driven cooling fan. (Courtesy of Bull Electric Ltd.)

that the same current flows both in the armature and field circuit (see Fig. 2.1).

Series connected motors possess a performance characteristic which renders them particularly suitable for certain applications where, for example, high starting torques are required. This form of motor tends to be inherently stable, i.e. the motor speed falls with increased load.

The use of series motors in variable speed applications is generally limited to those situations which can tolerate a high degree of speed variation with changes of load. This feature can, however, be effectively used in traction duties where a high initial torque is required for starting and breakaway from rest and for other applications where it is desirable that the relationship between torque demand and speed calls for a falling characteristic (see Fig. 2.2).

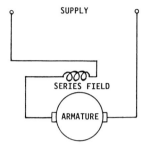

Fig. 2.1   Connections of a series wound d.c. motor.

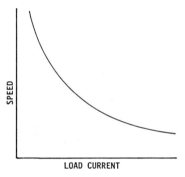

Fig. 2.2   Speed versus load characteristic of a series wound d.c. motor.

Shunt

Shunt motors have a field system connected in parallel with the armature circuit. In this design of motor the same applied voltage is impressed upon the field and armature. Variation of the applied voltage therefore affects both the field and armature circuits simultaneously (see Fig. 2.3).

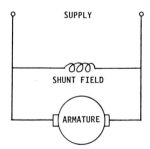

Fig. 2.3   Connections of a shunt-wound d.c. motor.

The basic relationships for speed, excitation and voltage apply to this type of machine and the operating characteristics and performance with variable voltage are given in Fig. 2.4. It should be noted that a given change in voltage applied to the motor does not produce a linear change in useful magnetizing flux and this, coupled with the demagnetizing effect of the current flowing in the armature, gives a naturally rising speed characteristic against increasing load (see Fig. 2.5).

The shunt motor possesses the ability to produce a sustained torque over a wide speed range with variation of speed and voltage. As this is probably the most frequently met requirement in industrial applications the shunt connected motor or a variation of it in the form of the separately excited motor is widely used.

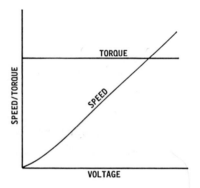

Fig. 2.4 Speed and torque versus voltage characteristic of a shunt-wound d.c. motor.

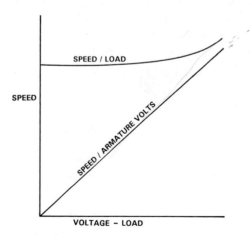

Fig. 2.5 Torque and voltage versus speed characteristic of a shunt-wound d.c. motor.

Separately excited

The separately excited motor behaves similarly to the shunt connected motor (see Fig. 2.6), and by deriving the excitation supply
from a separate source from that connected to the armature it is
possible to overcome some of the disadvantages of having to
design the field circuit for the same voltage as the armature. This
permits a greater choice in selection of winding materials and
thus contributes to a more efficient construction, giving the motor
designer greater flexibility in selecting the operating voltage for
the field circuit. The armature voltage is rarely an ideal choice for
this. The connections for a compound wound motor incorporating a
separately excited field are shown in Fig. 2.7.

The field system is effectively independent of the armature

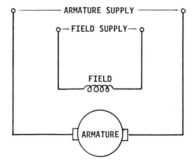

Fig. 2.6   Connections of a separately excited d.c. motor.

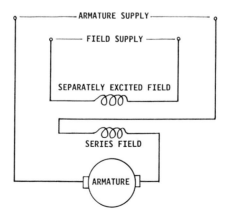

Fig. 2.7   Connections of a compound wound d.c. motor.

supply voltage and can be controlled entirely separately to give a motor of extreme flexibility in application. The motor can be designed to possess a stable characteristic (speed falls with applied load), although with modern systems employing armature voltage control and the growing demand for motors of ever greater efficiency, this feature is often sacrificed. The separately excited motor is the type of d.c. motor most frequently utilized in variable speed drive systems.

Compound

In their efforts to make d.c. motors more stable at all points of their operational range, the designers combine the inherently stable characteristic of the series connection and the speed/torque characteristic of the shunt connection. With this arrangement part of the excitation is derived from a winding connected in series with the armature and part from a separate winding connected either in parallel with the armature or from a separate source (see Fig. 2.7). By these means the motor can be designed with variable ratios of series and shunt winding to produce a motor with a characteristic eminently suited to many industrial applications. The effect of speed change with load can be minimized and the machine can also be made inherently stable.

Commutators and commutation

The commutator of a d.c. motor is generally regarded as the most questionable item where reliability and maintenance are concerned. However, properly designed motors, with commutators, working within their design parameters have been proved by experience to be reliable in practice. Evidence of this is seen in the exceptional life actually achieved in working environments, and many examples can be cited of d.c. motors, with their original commutators, in continuous operation for 20 to 30 years.

It is implicit in the function of a commutator that it is a component subject to wear, due to the rubbing action of the brushes, and for this reason it does require regular attention or at least regular inspection. Given proper conditions, good reliability can be expected provided that the brushes and brushgear have been correctly selected, designed and adjusted. The life of a commutator can be seriously reduced if excessive sparking occurs

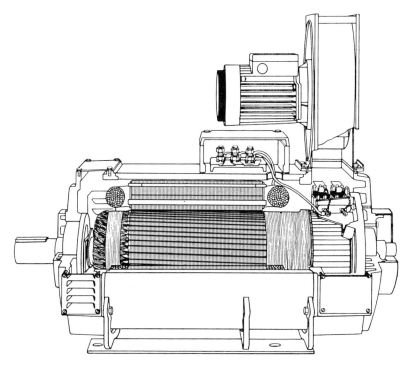

Fig. 2.8 Sectionalized drawing showing the detail of construction of a modern d.c. motor. (Courtesy of Thrige–Scott Ltd.)

and every effort should be made to reduce this. The degree of sparking can be controlled to a certain extent, by the user; however, in order to achieve this a basic understanding of the factors which influence good commutation is necessary.

The function of the commutator is to select the path of the energizing current through the armature winding so that a continuous unidirectional torque is achieved. In order to achieve this operation, which is essentially a switching process, the commutator and its associated brushgear progressively change the supply to the various armature winding elements as the armature rotates. By mounting the commutator on the rotating shaft and connecting the armature windings to it, the correct physical relationship between the flux developed by fixed field system and the flux developed in the armature winding is achieved when the commutator is supplied through the fixed brushgear.

Ideally the switching of the armature coils during the transition from one segment of the commutator to the next should be carried out when the voltage induced in the relevant armature coils by the action of the magnetic field is at or near zero. This is necessary because during the transition of the brushes from one coil to the next the coil is shorted out by the brush and if excessive voltages were present at the moment of commutation excessive current would flow across the surface of the brush, which has a relatively low impedance. The voltage induced in the armature coils is at its lowest point when the coil moves from an area where the magnetic flux reverses in polarity. This point is theoretically midway between the centres of adjacent poles.

In practice the precise location of the ideal point for location of the brushes or 'dead band', in a motor, is slightly behind (i.e. *against* the direction of rotation) the geometrical centre. This is due to the magnetic flux generated in the armature by the armature current interacting with the field flux; this effect is referred to as 'armature reaction'. The precise position of the 'dead band' will vary with changes of armature current. It is therefore normal practice for motor manufacturers to provide some means of adjusting the position of the brushgear in order to permit setting during construction; this also enables the user to obtain the best operational performance by making any necessary adjustments. Most suppliers of d.c. motors and generators set the position of the brushes for approximately 75% to 85% of full load unless otherwise instructed.

The use of interpoles on d.c. machines improves the flux distribution around the machine, and, as these are physically sited between the main poles and connected in series with the armature, the position of the 'black band' can be substantially maintained with varying loads. Total correction for all loading conditions between no-load and full-load cannot in practice be achieved and if the machine is required to work with wide load variations or to operate for short periods in overload it is important to ensure that the designers are made aware of this.

In considering the selection of type or 'grade' of brush for a given machine it is very desirable to be aware of the operating conditions and those factors which determine an ideal brush composition for those conditions.

The material utilized for brushes is usually a graphitic carbon composition which may contain other elements or metals, e.g. copper or silver. The 'ideal' brush material should have a low friction surface to avoid excessive wear to the copper commutator; and it should exhibit sufficient resistance in a circumferential direction to avoid excessive current flow between commutator segments during commutation. It is an unfortunate fact that the materials from which brushes are made do not naturally combine these features and some compromise has to be arrived at.

In specially designed motors for use in exceptional conditions careful selection of the brush grade is made in order to give the best possible results. For example, a machine operating in an exceptionally dry atmosphere − this is experienced at high altitudes where lack of atmospheric moisture contributes to a reduction of friction − should utilize a brush grade providing very good inherent lubrication properties. As carbon possesses a naturally high resistance, where low losses are important the brush composition incorporates greater quantities of metals.

To summarize, the design of a commutator and its associated brushgear is a complex process and the theory of the design takes into account many diverse factors quite apart from the environmental and applicational considerations. Among these are the contact resistance, linear speed, rotational speed, voltage to be commutated, inductance of the winding, current density and whether the design is to incorporate interpoles. (It should be noted that it is not usual modern practice to supply interpoles on d.c. machines of less than about 10 kW.)

An examination of the commutator of a machine after several hours of operation will usually indicate whether there is likely to be a problem in this area, and the following checks, whilst not being exhaustive, illustrate the salient points that should be observed.

- Excessive or aggressive sparking.
- Burning or pitting of commutator segments.
- Ridging of the commutator.
- Raw copper appearance.
- Formation of a hard glazed appearance.

The best commutator life is achieved when the area under the brushes exhibits a dark chocolate colour which is smooth and unridged and there is no sparking at the edges of the brushes. Minute multiple or pinpoint sparking is frequently observable, particularly at the trailing edges of the brushes; however, this is rarely harmful. Excessive sparking can occur for the following reasons:

- Vibration.
- High segments on the commutator.
- High slot insulation (caused by commutator wear or poorly carried out inter-segment insulation undercutting during maintenance).
- Pitting or the appearance of small nodules of copper, which can occur when the machine is subjected to excessive armature current.

A winding fault will usually result immediately in bad commutation with aggressive sparking; this condition must receive immediate attention. The design of some armature windings can give rise to unbalanced or uneven currents occurring within the armature, and this can be observed as a regular pattern-mark appearing on the commutator due to the unequal distribution of the current between the segments ('selective commutation'). This particular condition may also occur when a motor is required to operate for long periods at one speed and the supply has a regular pulsation.

A raw or 'just-machined' appearance is usually corrected by a change of brush grade as this is often the result of the brush material being too abrasive for the environment. Similarly the generation of a 'glazed' surface on the commutator, which exhibits a high electrical resistance, can usually be corrected by removal of the glaze by mechanical means (machining or application of an abrasive block), and then replacing the brushes with a more appropriate grade. In all cases where it is felt that commutator performance could be improved by a change of brush it is advisable to seek advice from specialist manufacturers. These companies have accumulated a vast store of empirical knowledge and have a programme of continuous development.

A.C. motors

Over the years many types of a.c. motor have been developed, some of which have evolved for specific applications demanding special characteristics. The a.c. motors most frequently met with in industrial applications can be broadly divided into the following types: induction, series commutation, synchronous, stator or rotor fed, pole-changing motors and reluctance motors.

Induction

The induction motor is the most widely used a.c. machine in power applications today. Operating from multi-phase supplies this type of motor is robust in construction and provides a useful torque characteristic (see Fig. 2.9), making it eminently suitable for a wide range of applications. Its use as a component of a variable speed system is limited by the fundamental performance of the machine as the speed of operation is dependent upon the applied frequency. Thus, in order to employ induction motors on applications requiring variable speed it is necessary to feed the motor with some form of variable frequency or to employ some means of changing the internal voltage balance. There are basically two designs of a.c. induction motor: the slip-ring and cage types. Both types have similar electrical characteristics and selection of either is largely a function of the starting requirements and possible limitations on starting torques and currents.

Cage

As the name implies the rotor winding of a cage machine is constructed from conductors made from bars of copper, aluminium or other alloys arranged around the rotor in the form of a cage. There is no electrical connection between the stator and rotor and the current which flows in the rotor cannot be changed arbitrarily from outside the motor. It is therefore only possible to alter the speed by changing the frequency.

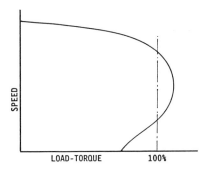

Fig. 2.9 Characteristic curve of speed versus load for a typical a.c. induction motor.

Plate 2.2   Rotor of a large modern cage-type motor in the course of production. (Courtesy of Laurence, Scott and Electromotors Ltd.)

Slip-ring

In the slip-ring motor the rotor is constructed with a winding which is brought out to slip-rings mounted on the motor shaft to which control gear can be connected to adjust the starting torque, current and, if required, the speed.

Both motors depend for their action on the magnetic field produced by the induced current in the rotor reacting with that generated by the current supplied to the stator. To achieve useful torque there has to be some difference in speed between the rotor and the rotational field developed by a multi-phase supply to the stator. This difference is termed the 'slip' and varies with load. In practice the value of slip is small, as maximum torque from the motor is achieved with only a 2−6% slip (see Fig. 2.10).

Series commutator

Series a.c. motors are connected in the way shown in Fig. 2.11. The commutator behaves like a switch which ensures that the flux in the rotor and stator is always acting so as to achieve a motoring torque. In this respect the a.c. commutator motor and the d.c. motor are similar in concept and for this reason the a.c. commutator motor is often referred to as a 'universal' machine and can

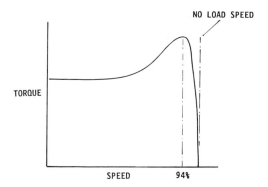

Fig. 2.10 Characteristic curve of speed versus torque for an a.c. induction motor showing change of speed (slip) with load.

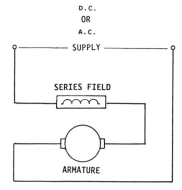

Fig. 2.11 Connections for a series wound a.c. motor.

be used on either a.c. or d.c. supplies. However, due to the different impedance of the windings when operating on a.c. or d.c., the supply voltages required are usually different. The design of this type of motor for use on a.c. supplies is difficult as it has to commutate or switch relatively high voltages giving rise to substantial sparking at the brushes. Thus motors of this design are usually limited in their application to smaller ratings and are chiefly applied in power tools and similar situations.

Synchronous

Synchronous motors by definition run at a speed directly pro-portional to the frequency applied. To achieve this the rotor is

Plate 2.3 Rotor of a large slow-speed salient pole synchronous induction motor.

polarized and is magnetically locked into the rotating field generated by the the a.c. supply. The motor differs from other forms of a.c. motor in so far as it does not possess an intrinsic starting torque when connected directly to an a.c. supply; therefore, a special induction motor type of winding or a small motor, usually short time rated, is introduced and coupled to the rotor in order to start the machine and get it up to a speed sufficiently close to synchronous speed to allow the rotor to 'lock' into the rotating field. Where used with variable frequency it is possible to start these machines by increasing the frequency from zero. A change in torque demand gives rise to small angular displacement of the rotor relative to the stator flux but there is no change in speed.

Stator and rotor fed

In all electric motors the voltage generated within the rotor winding must be balanced by the external voltage applied for it to operate at a stable and constant speed. In the case of cage type induction motors this voltage is that dropped across the impedance of the windings.

With slip-ring induction motors the balancing voltage is that dropped across the complete electrical circuit of the rotor which can include external impedances (Fig. 2.12). By adjusting the external impedance the current in the rotor circuit can be altered

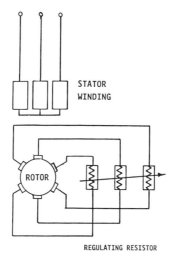

STATOR
WINDING

ROTOR

REGULATING RESISTOR

Fig. 2.12 Diagram of connections for a typical stator-fed a.c. commutator motor with speed regulating resistor.

and it will then supply more or less motoring torque, which will cause the speed to change until a new balancing situation is reached. This principle can be used for the purpose of obtaining speed variation in a slip-ring motor but the actual operating speed will be load dependent.

A unique design of motor utilizing this feature was developed in which, by connecting the rotor winding to a commutator, it is possible to change the voltage generated at the brushes and hence, by altering the voltage balance within the machine, produce a change in speed. This technique can be achieved by two methods.

Rotor fed

In the 'rotor-fed' design the commutator brushes are arranged so that they can be moved radially relative to the stator; the commutator can thus deliver variable excitation to a rotor control winding which either assists or detracts from the magnetic field produced by the main rotor winding, thus changing the voltage balance, and the speed alters until a new balancing condition has been met.

Stator fed

An alternative connection, the 'stator-fed' design, utilizes an external regulating system connected to the commutator to achieve the same effect; in this form of motor there is no requirement for a stator control winding as all the adjustment is made directly in the external regulator and fed to the commutator directly. A typical example of the form of motor which uses an external speed regulating system can be seen in the Laurence Scott N–S motor, the connections for which are shown in Fig. 2.13. This motor has been successfully employed on many large variable speed applications, some involving many hundreds or even thousands of kilowatts.

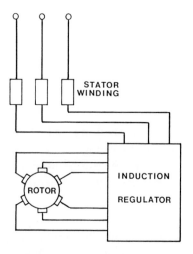

Fig. 2.13   Basic connections for a series connected stator-fed a.c. commutator motor with speed regulating induction regulator; this connection gives a characteristic similar to that of a series connected d.c. motor.

Motors which utilize brush shifting techniques are generally found in the smaller size of application, typically up to about 300 kW, due to design limitations. These motors tend to be physically large for a given output; this can preclude their use in situations where a very high speed of response is required and where a high inertia is undesirable. The connections for the rotor and stator of a stator-fed commutator motor are shown in Fig. 2.14.

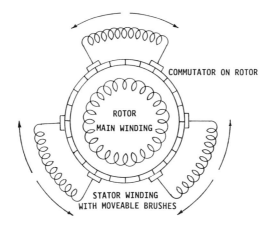

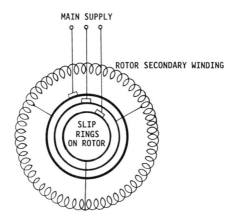

Fig. 2.14 Connection and disposition of windings in a rotor-fed commutator motor.

Pole changing

As indicated earlier, the fundamental speed of a.c. motors of the induction type is essentially a function of the applied frequency and the number of effective poles designed into the machine winding. It is possible to design and wind a motor in such a way that the number of poles can be changed by employing a number of windings with tappings or connections, so that reconnecting the stator winding changes the number of poles and hence gives variable discrete speeds, each speed being a function of the number of poles. The change in speed is a step function and is not a smooth continuous transition from each speed to the next. The nominal operating speed of the motor is thus changed in inverse proportion to the number of poles created, i.e. the smaller the number of poles the greater the operating speed. Whilst in theory there is no limit to the number of speeds that could be achieved by this method (the highest speed, of course, being that achievable from two poles – 3000 rpm for 50 Hz) practical considerations of space and efficiency normally limit this type of construction to two or possibly three speeds.

Pole-changing can also be obtained by connecting a single winding with tappings in such a manner as to create a variable number of virtual poles by modulating the flux around the stator. This technique is often descibed as 'pole amplitude modulation'. It is achieved by distributing the winding on the stator so that by connection of the supply at various points the number of poles can be changed (see Fig. 2.15). This motor differs from the conventional pole-changing motor in that it utilizes all the winding at all speeds of operation. Pole-changing motors are usually designed with a fairly close choice of operating speeds, e.g. 1000 and 1500 rpm, 750 and 1000 rpm or 600 and 750 rpm.

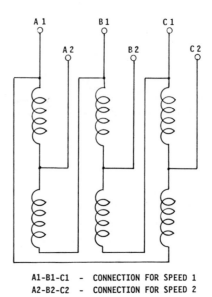

```
A1-B1-C1  -  CONNECTION FOR SPEED 1
A2-B2-C2  -  CONNECTION FOR SPEED 2
```

Fig. 2.15 Principle of connections for a pole amplitude modulated a.c. motor.

Reluctance

The reluctance motor depends for its action on the principle that, given the freedom to rotate, an iron rotor mounted within a magnetic field system will move so that it takes up a position in which the magnetic flux travels through a path of minimum reluctance, and in effect the rotor lines itself up with the lines of force of the field. If the field is now made to rotate the rotor will be dragged around to follow it.

Motors of this basic construction have been in existence for over 150 years but only relatively recently has the availability of solid state switching devices for generation of the rotating field made these motors a viable proposition. The early application of this type of motor was inhibited due to its inability to generate a starting torque when connected to a fixed frequency supply, and to its relatively low efficiency. The development of solid state switched fields has generated considerable interest in this type of motor and the switched reluctance drive (SRD) has become an established contender in the drives market. The rotor is extremely simple and, as it has no windings, it is both robust and light in weight; this makes it particularly suitable for many applications. The current level of design is limited to about 250 kW and the SRD has made most impact on the lower end of the medium size range of 1 to 80 kW.

Chapter 3
Open and closed loop control

The control of a process or a process variable can be effected by:

- Measuring and adjusting any changes that may occur in the elements of the system which forms part of the process.
- Monitoring the end product of the process.
- Controlling the process against an arbitrary, predetermined set of values.

The objective of all control systems is to achieve a desired performance by utilizing appropriate information relating to the behaviour of the complete system, or a critical part of it, and then applying any corrective action that may be necessary as a result of some event having caused a deviation from this performance. The information to be utilized can take several forms and might include data on the environmental conditions as well as measurement of the controlled function.

A system that attempts to predict a requirement for a change of state is called a 'feed forward' system, and this can be a very powerful control strategy if the precise information is available on the effects of a given disturbance. Such systems cannot, however, be pre-programmed to react to every possible event or to compensate for changes for which the effect is not possible to predict. For this reason 'feed forward' systems are rarely used in practice and 'open' or 'closed loop' control is mainly employed.

A system that employs information derived from the process or some other measurement of performance is said to be operating in closed loop control. Likewise a system that uses preset arbitrary values is operating in open loop control. The fundamental difference between the two systems is that in closed loop control there is feedback of information from the process or controlled item; this does not occur in an open loop control system.

Feedback

Feedback, as its name implies, is the information that is used by the controlling element of the system to maintain or adjust the controlled device or process in order to counteract any deviation from the desired performance. In the case of a drive motor the feedback may act to keep the motor speed at a constant desired setting while it is subjected to any events that may have a tendency to change it.

When the feedback acts to negate the potential effects of a change to the operating conditions it is called 'negative' feedback. 'Positive' feedback is rarely, if ever, used as it produces instability owing to its effect on the controller, which acts to increase the deviation from the required set point.

The information derived from the measurement of the process can be proportional or have some other relationship to the error. For example, measurement of the position of two items with respect to each other would produce a signal which would be proportional to the difference in position; this can be utilized to give a most precise form of control. The speed difference between the same two items could be derived from this information by determining the rate of change of position or, as is more usual, directly from a speed measuring device, and the acceleration could similarly be derived by calculating the rate of change of speed. When applied to rotating drive systems these control signals become shaft position (angular displacement), speed (revolutions per unit of time) and acceleration (change of speed per unit of time). When all of the above elements are utilized in the control system it is referred to as PID control (proportional, integral and differential).

In the case of a speed-controlled system which has a disturbance imposed on it the proportional error will increase to the value which corresponds to the speed error at a rate determined by the accelerative response of the system, whereas the integral term will continue to increase in value as long as there is a speed error and, after this, until the position error is reduced to zero. The differential (acceleration) term will only appear while the speed is actually changing. In practice a well designed system will utilize all three terms for fast and precise control.

The feedback can take the form of a continuous stream of information which is compared with the reference signal; it can also be discontinuous and in this type of control the feedback and referencing signals are periodically sampled at regular time intervals. The latter form of closed loop control is often described as 'pulsed data control'. A further example of discontinuous feedback is one in which the feedback control value reaches a prescribed level prior to the system reacting. This type of control is commonly called an 'on–off' or 'bang–bang' control.

Open loop control

A simple example of open loop control of speed would be a motor driving a pump, filling a tank supplying liquid to a process, where the control is a simple push-button-operated starter (see Fig. 3.1). On being started the motor will continue to drive the pump regardless of the level of the liquid in the tank.

Closed loop control

To change the system in this example to closed loop control a level detection switch could be added, arranged to stop the motor and hence the delivery of the pump on achieving the desired level of liquid. This application is an illustration of a very simple system and is an example of on-off control. To improve the control system and prevent the motor, in this example, from turning on and off repeatedly it would be appropriate to introduce another level detector switch set to detect some minimum level, with the circuit arranged so that the motor will only commence to

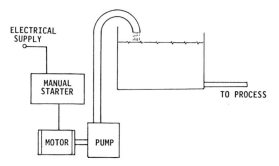

Fig. 3.1 Simple manual process control applied to a pumping application.

run again after reaching the lower level, and will stop only when the maximum level has been detected (see Fig. 3.2).

If the speed of the motor, and hence the delivery of the fluid from the pump, could be accurately maintained or measured and similarly if the flow from the tank could be measured, it would be possible to match exactly the speed of the motor and delivery of the pump so that after the tank had been filled a constant level could be maintained.

The modified control requires knowledge of the relationship between the motor speed and the pump delivery, i.e. the characteristic curve of the pump. This relationship is sometimes called the control algorithm of the system and is the mathematical expression relating one variable to another — in the above example, the pump speed and delivery from the pump.

Examples of control systems

Almost all industrial manufacturing activity incorporates process control utilizing feedback and closed loop control. A good example of this can be seen in the manufacture of paper. In this process fibres of wood or other materials, highly dispersed in water, are discharged onto a continuously rotating band constructed of mesh. This permits a large proportion of the water to drain away by gravity or by the application of vacuum; a residual sheet of enmeshed and evenly distributed fibres is thus formed. The sheet is passed into a series of rotary presses where more of the residual

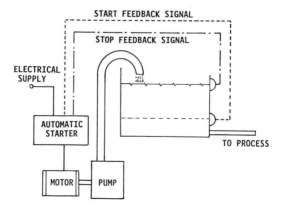

Fig. 3.2   Closed loop on/off control system applied to a pumping application with simple level control.

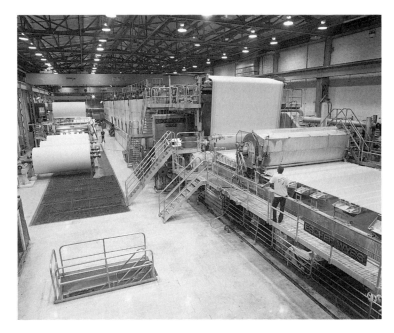

Plate 3.1 View of a large modern paper-making machine equipped with a sectional electric, multi-motor, digitally controlled thyristor d.c. drive. (Courtesy of Sittingbourne Paper Company Ltd.)

water is expressed and the sheet is consolidated. Reduction of the residual moisture to the required level is completed by pressing the sheet into close contact with large steam-heated cylinders. Finally the finished sheet is reeled up for transfer to other processing or for cutting up into marketable sizes.

The finished product is required to conform to very strict criteria and manufacture is closely monitored and controlled at each stage of this integrated continuous process to ensure compliance and acceptability to the end user. The specification to which paper is made includes, amongst other things, thickness (calliper), weight (basis weight or grammage), surface finish (printability), colour and strength. These factors can all be influenced by uncorrected variations in the process, e.g. fibre dilution (stock density), rate of discharge onto the mesh, linear pressure of the presses, vacuum levels, temperature of the drying cylinders and the tension in the sheet. These variations can be introduced by changes to the ambient conditions (temperature, atmospheric pressure, humidity, etc.), as well as mechanical variations within the machinery, e.g. changing friction loads. Supply conditions and quality of the raw materials, together with variations in the basic utilities (electricity, gas, water, etc.) may also necessitate compensatory adjustments to the process.

The designer of a process of this type seeks to identify and measure all the process variables and apply closed loop control where this is possible or practicable. Figure 3.3 shows a simplified paper-making process and identifies the feedback paths and the various control loops. It is of interest to note that the activity that is being controlled is often acting indirectly, or in conjunction with another activity, in response to a measured change in state of a monitored function. In the figure, the fibre dilution is controlled by both the quantity of water delivered (discharge rate of the pump) and the quantity of raw pulp delivered (conveyor speed). In order to design complex processes such as these correctly, the designer must be aware of the interaction of the

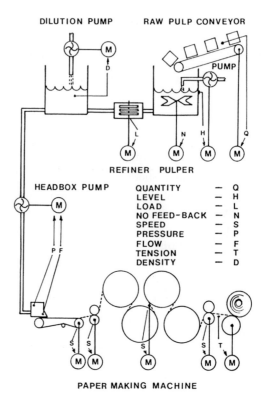

QUANTITY	—	Q
LEVEL	—	H
LOAD	—	L
NO FEED-BACK	—	N
SPEED	—	S
PRESSURE	—	P
FLOW	—	F
TENSION	—	T
DENSITY	—	D

PAPER MAKING MACHINE

Fig. 3.3   Simplified flow diagram of the process of the manufacture of paper showing the various feedback loops relevant to variable speed control.

various parts on one another as well as the basic control algorithm of the controlled item.

A typical control scheme for varying the speed of a motor forming part of the process described above is shown in Fig. 3.4. In this scheme the controller monitors the speed of a d.c. motor and adjusts the terminal voltage of the motor so as to keep the speed accurately following the speed demanded by the process. To detect the actual speed of the motor in this example use is made of a tachometer generator, which is a device that has an electrical output which is directly proportional to its speed of rotation. Such devices are designed to meet stringent specifications and typically the electrical output is proportional to speed with accuracies better than one part in one thousand or 0.1%.

Characteristics of control systems

Because of time delays in a control system it is not possible to make the corrective action of the speed adjustment exactly coincident with the onset of a speed change and therefore the correction is

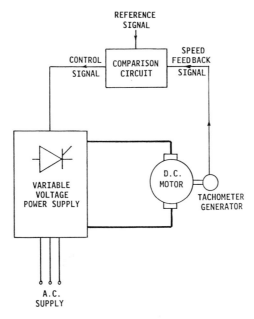

Fig. 3.4 Typical control scheme for a variable speed control system with closed loop velocity feedback control.

applied with some time delay which results in an oscillatory response. If the speed control feedback and the control system are properly designed the oscillation will quickly decay and a stable condition will be reached.

Examples of such a control system are shown in Fig. 3.5. The top diagram shows a system where the feedback is not properly adjusted and the oscillation will be maintained; this condition is called 'underdamped'. The centre diagram shows a system where the feedback circuit is correctly designed and this condition is called 'critically damped'. The bottom diagram is a situation where the feedback is adjusted to a too low value and this is called 'overdamped'. The ideal or critically damped condition is the response which gives the minimum departure from the desired or control level for the shortest time.

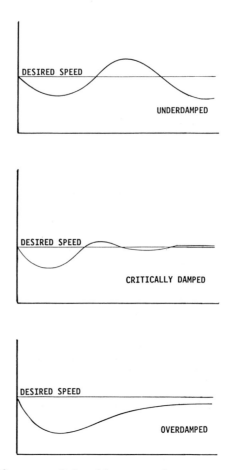

Fig. 3.5   Performance of closed loop control systems.

Open loop control of the pumping application shown in Fig. 3.1 behaves in the way shown in the top part of Fig. 3.6 where the speed of the system illustrated goes directly to the maximum followed by a period at rest. This would correspond to the system performance shown in the bottom part of Fig. 3.6.

The same system using closed loop control would perform in the way shown in Fig. 3.7.

To achieve satisfactory control from a system a number of procedures must be carried out, including the establishment of the functional requirements of the process control. This is called the systems specification and would include such data as the effect on the process of the control elements and the performance required from these elements to meet the process variables. This could typically be the accuracy required for the speed of a controlled motor, the input torque, the tension in a web of material, the delivery of a conveyor, etc.

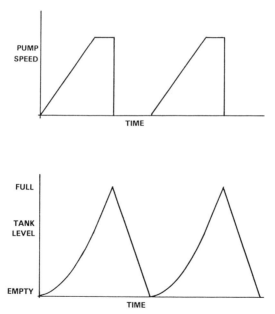

Fig. 3.6 Behaviour of a typical on/off control system, as applied to a pumping process.

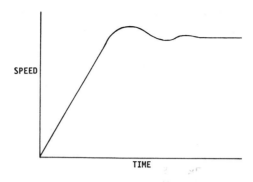

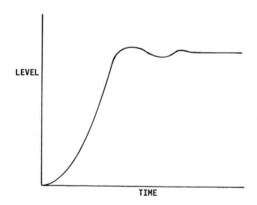

Fig. 3.7 Behaviour of a closed loop control system, as applied to a pumping application.

Performance specification

In order to specify the requirements of a control system to a potential supplier, manufacturer or designer of a control system a set of standard values are required. These standards of control are aimed at providing a basis for the establishment of performance and are usually given in the following form.

Steady state performance

This is generally expressed as the value at which the system settles after some disturbance and is often given as a percentage value of the preset or desired value to which the system is being controlled. Typically this value for a high performance drive system would be 0.01% of the maximum speed. For less critical applications the figure is relaxed to 0.1% although with modern drive technology it is unlikely that this will result in economies.

Transient performance

This defines the maximum excursion from the preset or controlled value following a system disturbance and is also defined as a percentage value. To completely specify this parameter it is also necessary to specify the nature and magnitude of the disturbance or disturbances to which the system is subjected as the transient departure from the preset speed is in most cases a function of both the magnitude and rate of application of the disturbance. A typical requirement would be a maximum departure from the preset speed of 0.25% for a load change of 10% of full load, applied as a step function when operating at a nominal load of between 10% and 80% of full load with a supply voltage variation of ±10% and a supply frequency variation of ±2 Hz. Additionally it is usual to specify a range of operating temperatures (typically 0°C to 40°C) and a temperature change rate of not more than 10°C in 30 minutes.

As indicated above, system disturbances can be the result of a number of imposed changes and these can be specified as occurring independently or simultaneously. The performance criteria may also include a requirement for a specific settling time after the onset of these disturbances. It can therefore be seen that if various systems are to be specified, evaluated or compared any definition of the performance must be accompanied by the conditions under which the performance is achieved.

The application of closed and open loop control

Variable speed drives, as employed extensively in industry, use both closed and open loop control. In those situations where accurate speed control is essential and wide speed ranges are required closed loop control is generally used.

Both a.c. and d.c. drives are available with closed loop control and if accurate speed control is required from an a.c. drive using conventional induction motors closed loop control is essential.

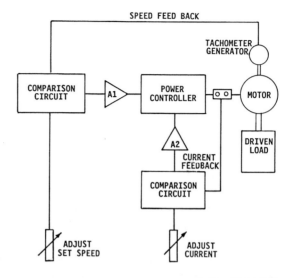

Fig. 3.8   Basic control circuit diagram for a speed-controlled drive system.

The majority of d.c. drives are designed with closed loop control as a standard feature.

Closing of the speed control loop on a drive system entails determining the actual speed of the driven motor, comparing this speed with the required control reference, determining any error and sending to the drive controller the necessary command to reduce the deviation from the preset speed to zero. The basic control diagram is shown in Fig. 3.8. In this circuit the feedback from the speed-sensing tachometer is compared with the speed demand derived from the speed setting circuit. Any difference in these values generates an error signal which is passed via an amplifier A1 to adjust the output from the power controller in such a way as to reduce this error to zero by altering the speed of the controlled motor. Control of the current/torque demand by the motor is achieved via the current feedback system and amplifier A2, which forms a separate control loop inside the velocity control loop, as shown in the figure.

Chapter 4
Power semiconductors

The term semiconductor is applied as a general description of a large number of electrical components in which a material or number of materials in combination behave electrically as a nonlinear impedance, i.e. the flow of current is not directly proportional to the voltage applied. The degree of nonlinearity to the flow of current can be adjusted in certain of these devices by the injection of small currents, and in this sense they can be used as amplifiers or switches in which small amounts of control power can be made to control or switch considerably larger power circuits.

The manufacture of the material from which semiconductors are made has to be carried out in a very carefully controlled environment as impurities can change considerably the characteristics. Semiconductors are used extensively in the electronics industry; the development of personal computers and miniaturized radios, for example, could only have been accomplished by the development of these components. The phenomenon of semiconduction has been known for many years and the ability of certain materials to behave in a nonlinear fashion has been exploited in the past for the construction of electrical surge suppression equipment and solid state rectifiers.

Typical of early semiconductors was the copper oxide and the selenium rectifier, and while these early devices did not possess the property of controllable nonlinearity they were extensively employed as solid state rectifying devices. The physical properties of semiconductors can be changed during the course of their manufacture by the amalgamation into their structure of minute amounts of modifying substances, and their manufacture is a very highly developed industry.

The semiconductors described below include the devices most often met with in power circuits.

Diodes

Diodes are two terminal solid state electronic devices which allow current to pass in one direction only and present an extremely high (almost infinite) resistance to current applied in the opposite direction, provided the voltage applied in reverse does not exceed the specified working level of the device. If therefore an alternating current is applied to such a device current will pass through in one direction only and thus the device rectifies the current (Fig. 4.1). A multiphase circuit is shown at Fig. 4.2.

Diodes can be constructed from a variety of base materials, the earliest ones were made from a combination of metals and oxides of metals, for example iron and copper oxide, selenium, germanium, etc. These early diodes have been largely superseded for power conversion purposes by diodes made from silicon. The

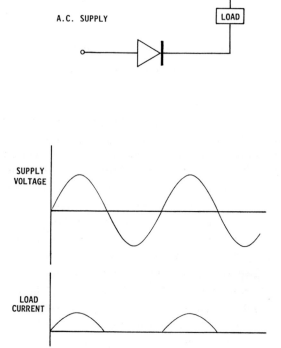

Fig. 4.1 Single-phase rectifier circuit and associated input and output voltage waveforms, illustrating blocking of reverse voltage in the negative half-cycle of the a.c. supply.

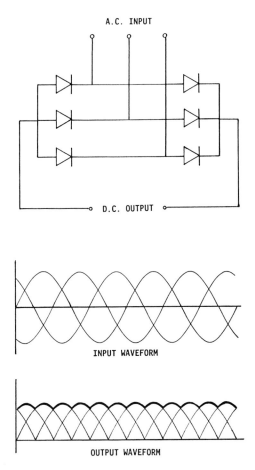

Fig. 4.2 Three-phase bridge-connected rectifier with input and output waveforms, illustrating how the effect of multiphase operation combined with full-wave rectification produces a much smoother d.c. voltage.

actual rectifying process takes place at the surface point of contact of the dissimilar metals or crystalline salts or oxides of metals, and a typical silicon device is shown in diagrammatic form in Fig. 4.3. In this diagram the two dissimilar materials labelled P and N have quite different atomic structures. In the case of the P material the bonding between the constituent atoms is incomplete and it can accept additional electrons; conversely the N material has loosely combined or 'free' electrons; these free electrons can be made, under the influence of an electrical potential, to migrate from the N material to the P material and current will flow through the device. If the electrical supply is reversed in direction

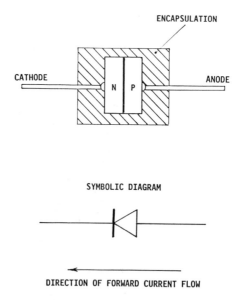

Fig. 4.3 Construction of a simple rectifier and symbolic diagram showing direction of forward current flow.

the free electrons are moved away from the junction area and the device exhibits a very high impedance to the flow of electrical current.

In order to block reverse current flow the device has to be able to resist the applied reverse voltage. The ability to withstand this voltage is described as the 'inverse voltage characteristic' and the maximum voltage that such a device can withstand is called the 'peak inverse voltage' (PIV) rating of the device, to distinguish it from the operating, or rms, voltage.

Silicon diodes are designed and manufactured for a wide field of applications ranging from fractions to many thousands of amperes; they can be obtained with operating voltages in excess of 4000 V. These devices are ideally suited for large-scale rectification systems as well as the multitude of applications that require the simple uncontrolled conversion of a.c. to d.c. They are used extensively in the first stage of inverters for power supplies and in traction duty.

Transistors

Transistors are also semiconducting devices insofar as they allow

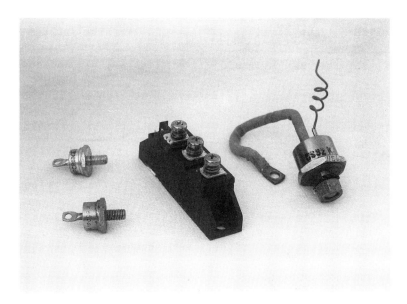

Plate 4.1 A selection of semiconductor devices – diodes and thyristors of the type used in small variable speed drives.

current to pass in only one direction, but they also possess a property that enables them to change the impedance to the forward flow of current by the injection of a small control current into another connection near to the contact point in the device; they are therefore three-terminal devices. Of the three connections, the emitter (E) and the collector (C) comprise the main current-carrying connections, while the controlling connection is made at the base (B). Diagrams of typical transistors are shown in Fig. 4.4.

The small controlling current which flows in the base circuit directly affects the forward impedance and can therefore be used to control the main current flow between the emitter and collector. Thus it can be seen that the device can be used for control or amplification as the forward current is proportional to the change in the small current used to control the forward impedance. Typically, a control current of a few milliamperes can be made to adjust a load current of fifty or more times this value which, in addition to making this an excellent and efficient amplifier, enables it to be used as a solid state switching device.

The transistor has undergone steady development since its emergence as a low-voltage, low-power amplifying device, and considerable research has been undertaken in order to improve

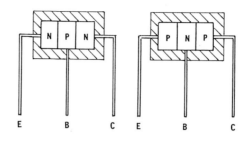

Fig. 4.4   Typical transistors of the PNP and NPN type showing basic construction and symbolic diagrams.

its performance as well as to extend its range. Transistors are now available for use in power circuits with ratings of up to several hundred amperes and for operation at up to 200 or 300 V. Arising recently from these developments has been the insulated gate bi-polar transistor (IGBT) which has the ability to control currents of up to 800 amperes at operating voltages in the region of 800–1000 V. It is generally accepted that the IGBT will in future occupy a considerable area of application currently filled by the gate turn-off thyristor (GTO).

Thyristors

Thyristors were developed from the silicon diode and thus possess the latter's ability to allow current to pass in one direction only. The thyristor, however, is provided with a control electrode similar to the transistor and is therefore also a three-terminal device. The thyristor differs from the transistor in that, when current is applied to the controlling electrode of a thyristor, the device goes into a freely forward conducting mode which cannot be changed until the current flowing falls to zero. If an alternating current is applied the thyristor can be made to conduct for only part of the cycle. The physical arrangement of a thyristor is shown in Fig. 4.5

Plate 4.2 A power thyristor with clamping device for fastening to a cooling-fin or other form of heat exchanger.

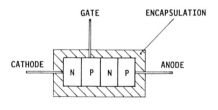

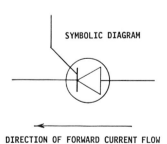

Fig. 4.5 Construction of a typical silicon-controlled rectifier (SCR) or thyristor with symbolic diagram showing direction of forward current flow when in the conducting mode.

and an example of the way in which this device operates in an alternating current circuit is shown in Fig. 4.6.

Gate turn-off devices (GTOs)

A relatively recent development of the thyristor is a device commonly described as the 'gate turn-off thyristor' (GTO). In this form of semiconductor a controlling circuit has been incorporated which permits the device to be restored to its non-conducting state after being induced to conduct. The thyristor does not possess this property and once in a conducting mode continues until the current falls to zero. With a GTO it is possible to turn the device off without waiting for the current to fall to zero by the application of a gating pulse.

In many respects this makes the device behave like a transistor

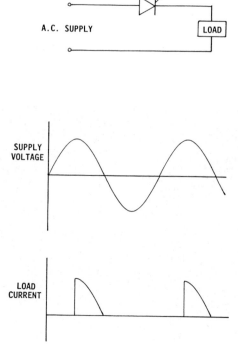

Fig. 4.6 Rectification and control of a single-phase a.c. supply by means of a thyristor. The thyristor control circuit is operating so as to permit conduction to commence at approximately 90°.

but unlike the transistor this is not an infinitely variable change in forward impedance but a straightforward on or off condition of conduction. The range of the GTO is probably the widest of all the controlled semiconductors, having been extended up to 4000 A. These devices are used extensively in high power inverting circuits and for traction duty.

To effectively 'turn-off' the forward current of GTOs specially designed circuits have been devised; these circuits require considerably more power than that used for switching thyristors and for this reason GTOs are less efficient. The amount of power used in the control circuits is not directly proportional to the size or forward capacity of the device and therefore the power absorbed in the control circuits has less significance in the larger sizes, which is why their use is chiefly confined to the higher power applications.

Protection of semiconductors and loads

The conducting area within the semiconductor is usually very small at the onset of conduction, but it rapidly spreads over the surface of the device. During the process of the propagation of this conducting area the device is prone to breakdown if subjected

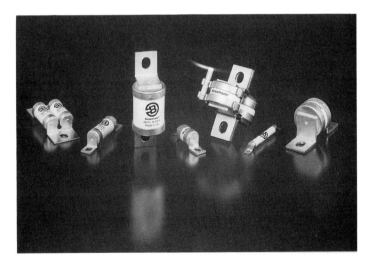

Plate 4.3 A range of fuses with specially designed characteristics for the protection of power semiconductors. (Courtesy of Bussmann Division – Cooper (UK) Ltd.)

to too rapid an increase in current. To protect these devices it is necessary to ensure that the rate of rise of current is limited to that appropriate to the design. This is usually achieved by inserting 'rate of rise of current' limiting reactors when the load does not possess sufficient reactance to achieve this. Semiconductors are also sensitive to rapid rise of voltage across the device which can

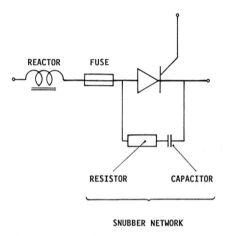

Fig. 4.7 Typical connections of protection devices used in conjunction with semiconductors.

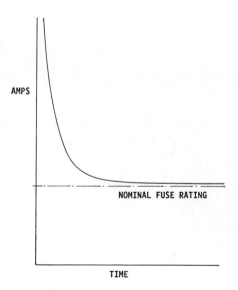

Fig. 4.8 Characteristic curve of fault current versus time for a semiconductor protection fuse.

also cause breakdown; protection against this is achieved by connecting, in parallel with the device, a circuit composed of resistance and capacitance to conduct excessively fast changing voltages away from the device; this circuit is termed a 'snubber network'. All these protective devices are shown in a typical circuit configuration in Fig. 4.7.

To limit overcurrent a range of special fuses has been developed. These fuses have very fast operating characteristics and can be selected to give a high degree of protection. A typical characteristic curve of a semiconductor fuse is shown in Fig. 4.8 and a typical construction of these fuses is shown in Fig. 4.9. This type of highly specialized fuse is designed to break the circuit rapidly when subjected to a fast rising surge of current, and thus protect the semiconductors which are very susceptible to such surges.

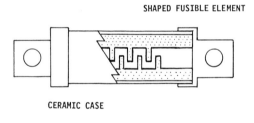

Fig. 4.9 Typical construction of a rapid-acting thyristor protection fuse.

Conventional protection equipment is often incorporated, specifically designed to protect the loads fed from thyristor systems; and these can typically comprise thermal or magnetic overloads.

As additional protection the cooling fins of the semiconductor devices are often fitted with thermostats or other over-temperature sensing devices designed to interrupt the flow of power through the thyristors in the event of overheating.

Semiconductors in inverters

Circuits designed to convert d.c. to a.c. are termed inverters, and they are generally used to produce variable frequency from a d.c. supply. The basic principle commonly employed is to fully convert the supply into d.c. and then divide the d.c. into a series of pulses of either varying width or amplitude. The resulting waveform can therefore be either of a stepped (see Fig. 4.10) or variably spaced form (see Fig. 4.11).

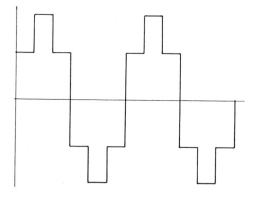

Fig. 4.10   Ouput waveform of a six-step inverter simulating a sine-wave.

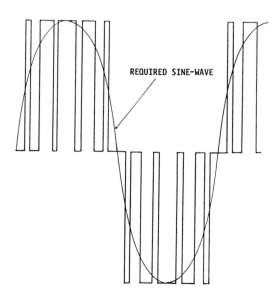

REQUIRED SINE-WAVE

Fig. 4.11   Output waveform from a typical pulse-width-modulated inverter simulating a sine-wave.

The typical connections of semiconductors in an inverter configuration are shown in Fig. 4.12. It will be noticed that this is similar in many ways to the converter with the addition of an incoming rectifier bridge circuit. To ensure that the current falls to zero (a prerequisite for the thyristors employed in this form of circuit) a special circuit is employed which ensures that the devices are able to switch off and on as required. Inverters can be constructed from transistor devices and in the larger sizes from gate turn-off thyristors. In principle, however, the basic power circuit remains the same.

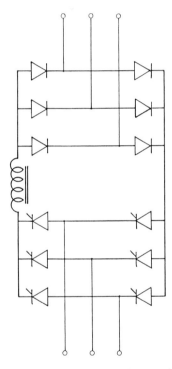

Fig. 4.12 Three-phase inverter circuit with a rectifier first stage, a thyristor controlled output stage and d.c. link reactor.

Semiconductors in converters

Circuits designed to change a.c. to d.c. are called converters and can be designed to work from single- or multiphase supplies. A single-phase converter comprises in its simplest form a diode connected in series with the load and the supply (see Fig. 4.1). The current is discontinuous because as the polarity of the supply changes so the diode successively permits current to flow and then acts to block it. Control of the current in the forward, or conducting direction, can be exercised by using a transistor type of circuit (Fig. 4.13) or a thyristor (Fig. 4.14). The discontinuity in the current gives rise to a strong pulsation in the flow of power and the maximum average voltage available from such a circuit is only 0.41 of the peak value of the supply voltage or 0.55 of the rms value. This is termed a half-wave rectifying circuit.

An improved and more frequently encountered circuit is that of the full-wave rectifier (Figs 4.15 and 4.16) and in this connection the current is sustained during both the forward and reverse voltage cycles although of course it does fall to zero as the

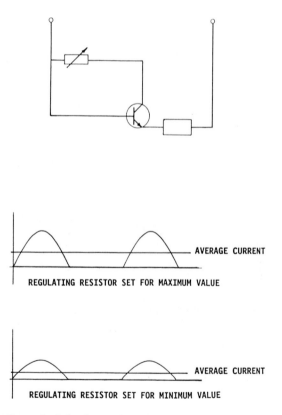

AVERAGE CURRENT

REGULATING RESISTOR SET FOR MAXIMUM VALUE

AVERAGE CURRENT

REGULATING RESISTOR SET FOR MINIMUM VALUE

Fig. 4.13 Control of the forward current in a transistor circuit showing connections and the effect on the forward current of varying the control current.

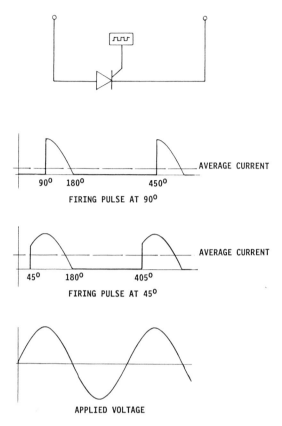

Fig. 4.14 Control of the average forward current in a single-phase thyristor circuit showing the effect of changing the firing angle.

voltage wave reverses. This gives rise to a smoother power flow and the maximum average voltage obtainable is equal to 83% of the peak value of the supply voltage, or 1.1 times the rms value.

Multiphase converters can be designed using both full- and half-wave connections and various examples of such circuits are shown in Figs 6.8 and 6.10. The current waveforms from multiphase converters are preferred for the larger applications involving variable speed drives.

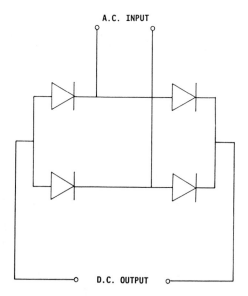

Fig. 4.15 Full-wave, single-phase, bridge-connected rectifier.

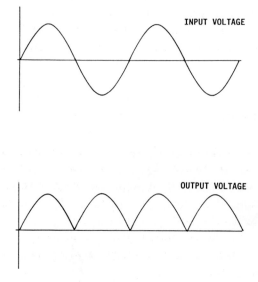

Fig. 4.16 Input and output voltage waveforms derived from a full-wave, single-phase, bridge-connected rectifier.

Chapter 5
Variable speed using a.c. motors

Historically the requirement for variable speed in power applications was met in three ways: by d.c. motors, by a.c. fixed speed motors combined with variable ratio mechanical or hydraulic devices, and by specially designed variable speed a.c. motors. Because of relatively recent developments in electrical circuit technology and the emergence of solid state control devices with the ability to handle substantial power (thyristors and power transistors) it is possible to design at reasonable cost and with acceptable efficiencies variable frequency supplies which permit the employment of induction motors for variable speed applications. The provision of variable frequency for this duty is by no means a new concept and rotating variable speed/variable frequency alternating sets were used many years ago for exactly this purpose.

Induction motors

The use of fixed speed standard induction motors is so widespread that it is obviously attractive to make use of this type of machine for variable speed applications when possible. But because, traditionally, the d.c. motor was used for variable speed duties and the a.c. motor principally in fixed speed applications, the extensive consideration given to the various aspects of d.c. motor operation over a speed range was not accorded to its a.c. counterpart.

One particular problem that has to be solved when using a.c. motors is the effect of the reduction in cooling air available at low speeds. Fixed speed motors are usually designed to be self-cooled, utilizing shaft-mounted fans, which blow external air either through ducts in the motor and/or over the motor carcase which is designed to provide the necessary heat exchange to dissipate the heat generated in the motor by the electrical losses. As the effectiveness of such a fan is directly related to the speed of the motor, in variable speed applications the cooling requirements may have to be accommodated by being either derived from a separate source or by heavily derating the motor and thus enabling the heat to be dissipated by surface cooling.

The method selected in practice is dependent on the load characteristics and the actual losses which occur over the speed range of the motor. Similarly, the starting torque and overload capacity available from a.c. motors has to receive careful consideration when these motors are used in variable speed applications and the fundamental characteristics must be always borne in mind.

The torque that an induction motor can develop is related to a number of factors that are implicit in this type of motor and its design and can be illustrated by the equation for the torque of the induction motor.

$$\text{Torque } \alpha \left(\frac{S \times R}{R \times (S \times X)} \right)$$

Plate 5.1 An installation of variable speed a.c. generators driven by N−S type motors, each providing variable frequency speed control for a group of a.c. motors. (Courtesy of Laurence, Scott and Electromotors Ltd.)

where: S = fractional slip
   R = rotor resistance
   X = leakage reactance of rotor at zero speed

From the above it can be seen that the reactance and resistance of these machines determines the useful torque that is available. For small values of slip the torque is approximately proportional to slip and for larger values of slip (over 2%) the torque approximates to being inversely proportional to the slip. The starting torque can therefore be low when compared to the torque available at maximum speed in a conventionally designed induction motor. This can be improved by specially designing the rotor to have a high resistance, by careful shaping of the rotor bars or by incorporating a double cage winding, one of which provides the main element of the starting torque, the other being used to provide the chief component of the running torque.

Improvement of starting torque by these means is only achieved at the expense of efficiency as the rotor loss is increased. It should also be noted that the maximum starting torque for which an induction motor can be designed will be the maximum torque that an induction motor can develop at any speed for a given set of supply conditions and is referred to as the 'pull out' torque.

Selection of a suitable motor for onerous starting conditions, therefore, devolves on either a specially designed machine or derating. Starting induction motors on fixed voltage and frequency supplies has to be done with care as the inrush current can be quite large (up to five times the full load current when direct-on-line started) and sustained when accelerating high inertia loads.

In those cases where undue stress would be placed on a supply system by direct-on-line starting, recourse is made to voltage reduction via a series reactor or a transformer, by successively connecting the motor in star and then delta or by adopting a technique known as 'soft starting'. Soft starters employ controlled semiconductors which control the current drawn by the motor at an appropriate level during acceleration, thus permitting smooth acceleration. This method of starting, which of course can be utilized also in variable speed applications where the motor is fed with variable frequency and voltage, provides the designer with additional freedom to optimize his selection.

The characteristics of the induction motor also make it necessary for care to be exercised in the selection of a machine that is required to operate with peaks of torque. In these cases the choice of a suitable size of motor is particularly necessary as operation at or near to the pull-out torque could result in the motor stalling.

Pole changing

Pole changing is probably the simplest form of achieving more than one speed from an a.c. motor. It does not have the ability to change smoothly from one speed to another and is therefore limited in the type of duty to which it can be applied.

The principle of pole changing derives from the basic relationship in an induction motor between the number of poles and its operating speed.

$$\text{Synchronous speed (rpm)} = \frac{\text{Frequency} \times 60}{\text{Number of pole pairs}}$$

To change from one speed to another it is necessary to create in the stator winding an alteration in the number of effective poles and this is achieved by a winding with appropriate tappings or by having two separate windings on the same stator. It is not usually possible to change from one set of windings to the other without disconnection of the motor during the transition.

Series resistance control

Rotor resistance control of slip-ring induction motors is a quite commonly used method of achieving a variable speed but it is inefficient over wide speed ranges and the actual operating speed is torque-dependent as the voltage appearing in the rotor is balanced by the voltage appearing across the regulating resistance which varies with the torque/load. Curves illustrating the change in speed with change in rotor resistance and load torque are shown in Fig. 5.1.

Specially designed a.c. motors

For many years it has been possible to obtain variable speed a.c. motors which, when employing specially designed windings

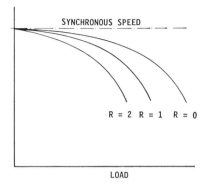

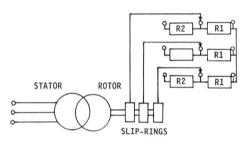

Fig. 5.1 Speed versus torque characteristic of a slip-ring induction motor illustrating the effect of varying the rotor circuit resistance.

and/or, regulating devices, can achieve a variable speed. A description of such motors has been given in Chapter 2.

These a.c. motors can provide an excellent method of speed control and have found extensive use in pumping and ventilation (fan) applications. Of the several varieties of specially designed variable speed a.c. motors the type to have received most attention is that which employs a commutator.

To achieve a stable speed in all electric motors the voltage generated in the rotor must be balanced by the voltage presented to the rotor by the external circuit. As the voltage generated in the rotor is determined by the flux and the speed of rotation, it is obvious that to achieve variable speed it is necessary to change the flux in the motor. This is achieved in commutator motors by moving the brush position on the commutator (Schrage method) or by using an external rotor voltage regulating system (stator-fed or N–S type). Control of speed with stator-fed a.c. motors has

Plate 5.2   Variable speed commutator motor of the N—S type driving a rotary cement kiln. (Courtesy of Laurence, Scott and Electromotors Ltd.)

probably received the most attention in recent years and it has met with considerable success. Control of speed to within 0.5% is obtainable with electronic systems which are used to adjust the regulating controller. Diagrams of the basic circuits of these motors are shown in Figs 5.2 and 5.3.

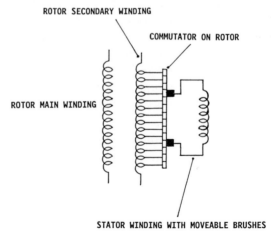

Fig. 5.2   Connection diagram illustrating the principle of control of a rotor-fed variable speed a.c. commutator motor with movable brushes.

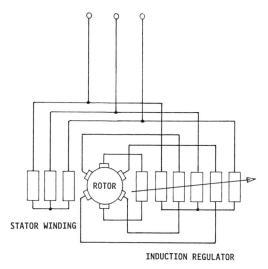

ROTOR

STATOR WINDING

INDUCTION REGULATOR

Fig. 5.3 Basic connections of a shunt speed characteristic stator-fed variable speed a.c. commutator motor (N−S type) with induction regulator control.

Slip recovery

As mentioned earlier, the slip-ring induction motor can be used as a variable speed motor by connecting the rotor to a variable resistance. The slip energy, that is the energy utilized in passing current through the resistance, is wasted in the form of heat. An ingenious method of recovering this 'lost' energy has been employed to improve the efficiency of this type of speed control, which is often referred to as a 'Kramer' system; the objective of the control system is to return as much as possible of the slip energy to the supply system.

In its originally conceived form the slip recovery induction motor employed a rectifying circuit connected to the slip-rings, which then passed the slip energy to a d.c. motor. This motor is mechanically coupled to the same shaft as the main motor. By adjustment of the field of the d.c. motor its back e.m.f. can be changed so that more or less slip power is absorbed by the motor and thus the speed of operation is changed. This energy is converted in the motor into useful energy which is added to the power delivered by the slip-ring motor. To achieve wide speed ranges and to cater for the conditions operating at starting, the rectifying circuits and d.c. motor have to be rated for the maximum voltage (which occurs at starting, i.e. the maximum slip position) and the current corresponding to maximum output.

For starting one of these machines it is usual to employ a conventional rotor resistance type starter which, having completed its starting cycle, is switched out and the converter switched into control at some pre-determined design speed of operation. Kramer drives in general are used mainly for relatively narrow speed ranges and are extensively utilized in pumping and fan applications where large powers are required. Figure 5.4 shows a simplified schematic diagram of the connections.

Static Kramer system

The principle of operation of the static Kramer drive is identical to the rotating system described above but in this case the d.c. motor is replaced with an inverter that converts the d.c. derived from the rectifier connected to the rotor into a.c. of the correct frequency so that the energy can be returned directly into the power supply (see Fig. 5.5). This type of drive can be designed to obtain wider speed ranges, typically up to 10:1.

The rating of the converter/inverter equipment is directly related to the speed range and load torque/speed characteristic and thus

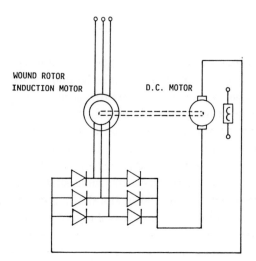

Fig. 5.4 Kramer slip power recovery, speed control system using a rectifier and mechanically coupled d.c. motor to return the slip power directly to the motor shaft.

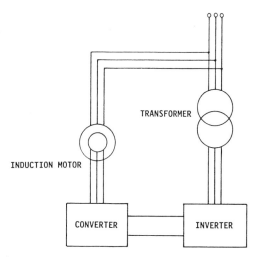

Fig. 5.5 Static Kramer slip power recovery, speed control system using a rectifier and inverter returning power to the electrical supply.

if the speed range can be limited there is a substantial saving in the size of this equipment. One advantage of the static slip recovery drive system over rotating systems is that the motor can be made to run at speeds in excess of the maximum 'natural' speed (the speed determined by the number of poles and the frequency), by injecting power into the rotor from the inverter.

Variable frequency control

All motors designed for operation on a.c. supplies can be made to operate over a variable speed range if a source of variable frequency can be provided. In the early days this was achieved by utilizing rotating variable speed alternators which were usually driven by d.c. motors which in turn received their supply from a d.c. generator driven at constant speed (see Fig. 5.6). Modern variable speed variable frequency drives employ solid state inverter circuits which are able to change the fixed frequency of the standard a.c. supply to variable frequency for feeding to the a.c. drive motor or group of motors.

The various circuits commonly employed for achieving this inversion, with one exception, have, as a first stage, a d.c. rectifier or supply. This supply is changed into a.c. by a process of

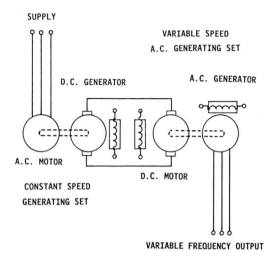

Fig. 5.6   Rotating system for providing variable frequency from a fixed frequency supply.

permitting current to flow in a carefully contrived pattern which is made to simulate a.c. By changing the pattern of the conducting and non-conducting cycles of the switching devices, the simulated a.c. can be changed in frequency. The exception to this method of achieving variable frequency is the method employed in the cyclo-converter (Figs 5.7a and 5.7b) which does not require a d.c. supply.

As the impedance of a circuit to an a.c. current varies with the frequency it is necessary, for proper control, to be able to adjust the applied voltage to prevent the flow of excessive current when the frequency is lowered and reduction of the current below that required to achieve the correct torque when operating at the higher frequencies. As the impedance of a motor is largely inductive designers usually arrange for the ratio of frequency to voltage to be constant.

Voltage-fed inverter

This term is applied to inverters in which the voltage applied to the motor is in general terms independent of the load being supplied by the motor. With this form of inverter the rectifier which forms the first part of the circuit is controlled in such a way that the d.c. output voltage to the inverting circuit is adjusted in accordance with the frequency demand.

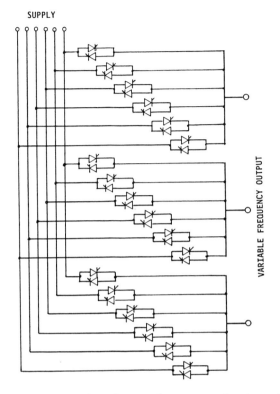

Fig. 5.7a Power connections for a cyclo-converter for three-phase, six-pulse operation.

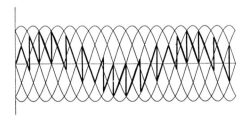

Fig. 5.7b Construction of a variable frequency simulated sine-wave obtained from a six-pulse cyclo-converter.

Current-fed inverter

This differs from the voltage-fed inverter in that the voltage output from the rectifier circuit is controlled generally in accordance with the load demanded by the controlled circuit or motor.

Quasi square-wave inverter

This form of inverter can be either voltage- or current-fed and the output waveform consists of an alternating voltage or current built up from a series of blocks of d.c. voltage or current of changing magnitude and polarity (see Fig. 5.8). In this type of inverter it is usual for the output waveform to be made up of six steps for each complete cycle. The frequency of the output is controlled by varying the timing of the conduction of the thyristors. In most inverting circuits harmonics are present due to the switching effect of the rectifying and inverting circuits. In a six-step converter the third harmonic and its multiples (6−9−12, etc.) are theoretically absent. By increasing the number of steps the harmonic levels can be considerably reduced.

In practice the higher the number of steps used to produce the simulated wave the greater the complexity of the circuit. It is not,

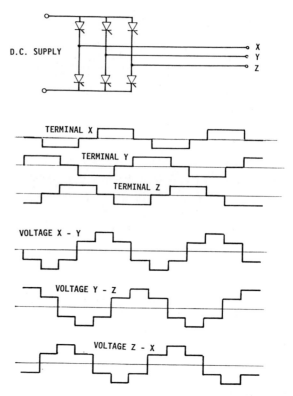

Fig. 5.8 Circuit diagram of the power connections and the waveform derived from a six-step quasi square-wave inverter.

therefore, economical to use more than six steps except for converters of higher power than about 100 kW. Exceptionally wide speed ranges are rarely satisfied with stepped type inverters due to the torque oscillations that may be generated by the motor reacting to the harmonic frequencies supplied by the inverter at low speeds. This torque oscillation can sometimes coincide with the natural frequency of the driven equipment and create a condition of mechanical resonance which gives rise to heavy vibration or mechanical failures. The power factor of stepped wave inverters is dependent upon speed of operation.

Pulse width modulation

With this form of circuit a d.c. supply is converted (or modulated) into a high frequency (up to 10 kHz) wave of square pulses of constant height and variable width. The required output a.c. waveform is then constructed from these pulses which are of very short duration (see Chapter 4, Fig. 4.11). This form of inverter gives rise to fewer harmonics. The high frequency of the basic intermediate frequency is considerably attenuated in the inductance of the driven motor and a relatively smooth approximation to a sine-wave is achievable. In some methods of construction series inductance may be added to improve the wave shape and further reduce the ripple. Pulse width modulated inverters can be fed from a simple diode bridge circuit of the unregulated type, as the voltage−to−frequency ratio can be satisfactorily adjusted in the modulation circuits. If the intermediate frequency is of a suitably high value, satisfactory operation can be achieved down to effectively zero frequency.

In some designs for drives with exceptionally wide speed ranges of operation, the modulation frequency may be changed several times throughout the speed range; this avoids the utilization of exceptionally high switching speeds at higher frequencies whilst still providing a sufficiently high modulating frequency for overall operation at very low speeds. This control design is sometimes referred to as 'gear changing'. An advantage of this form of a.c. inverter is that it operates at a very high power-factor which is independent of operating speed.

Cyclo-converters

Cyclo-converters differ fundamentally from inverters previously

Plate 5.3 A range of modern pulse-width-modulated variable speed a.c. drive modules incorporating the facilities for serial communication with a plant controller and other advanced features. (Courtesy of Cegelec Industrial Controls Ltd.)

described in that the a.c. supply is directly converted into an a.c. supply of a different frequency. This is achieved by switching the a.c. multi-pulse supply in such a way as to generate an average waveform which approximates to a sine-wave. The circuit shown in Fig. 5.7 is for a three-phase system from which it will be seen that, by using the thyristors to switch the output from phase to phase in a given sequence, the output terminals can be connected in turn to any of the input terminals. By selectively commutating (or switching) an output waveform can be constructed from parts of each phase of the incoming supply and this can be made to produce a multiphase supply. By varying the conduction times for each phase of the incoming supply the frequency can be varied.

In practice it is usual for cyclo-converters to be constructed from six, twelve or even higher phase circuits to achieve a better waveform at the output. It will be noticed that for a three-phase system 18 thyristors are used and for a six-phase system 36 thyristors are necessary. Two thyristors are necessary for each switching function so that current is able to flow in either direction.

The cyclo-converter cannot operate in its basic conception at frequencies much above half the supply frequency. This type of converter can operate quite successfully down to zero frequency but the usual practical limit for maximum frequency is normally set at between 20 and 30 Hz. It is possible, by utilizing a more complex circuit and/or special semiconductor devices to 'force commutate' (i.e. force the devices into non-conduction) and thereby achieve higher frequencies. This technique is only used commercially in the largest applications where the additional cost of the devices and controlling circuits is of less consequence. A disadvantage of forced commutation to achieve higher frequencies from a cyclo-converter is that the harmonics generated in the supply are increased and the efficiency is reduced.

The power factor of a cyclo-converter on six-phase operation is rather low, except at or near to the maximum frequency output; however, the harmonics generated by a cyclo-converter are generally lower than other forms of a.c. drive. As the cyclo-converter changes the supply frequency directly to the desired frequency the efficiency is good compared with the six-step and pulse-width-modulating inverters and the circuits which are used to turn on and off the commutating thyristors are simpler.

Chapter 6
Variable speed using d.c. motors

The speed of a d.c. motor on fixed voltage supplies can be varied by two methods, field control or armature resistance control.

Field control

Variation of the excitation (field) current of a d.c. motor will produce a change in speed. The fundamental relationship is that with constant applied armature voltage the speed is inversely proportional to the excitation flux (Φ).

Symbolically $\qquad N \, \alpha \, \dfrac{1}{\Phi}$

From this relationship it can be seen that for any given armature voltage the speed will be proportional to the field flux.

This principle of control has been widely used for many applications among them traction and mine winding – and also as an additional or trimming control where the main speed variation has been effected by change of armature voltage. In practice wide speed variation by means of this type of control is seldom attempted, because the field system becomes so weak that the motor may become unstable when subjected to high current variations due to the effect of armature reaction. This effect is illustrated in Fig. 6.1, which shows the behaviour of a d.c. motor when operating over varying load, and Fig. 6.2, which shows the performance of a similar motor with constant load and varying field.

Speed control by means of field adjustment is therefore limited in its application to situations where the range required is in the order of $2:1$ or $3:1$. Specially designed motors are used in certain circumstances where excessively wide speed ranges by

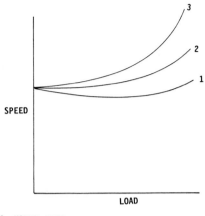

1 NORMAL FLUX
3 AND 2 WEAKENED FLUX

Fig. 6.1 Speed/load curve of a d.c. shunt connected motor with varying flux densities, illustrating the effect of armature reaction.

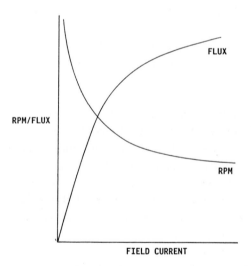

Fig. 6.2 Speed and flux density versus excitation current curves for a d.c. shunt motor showing the effect of changing the excitation level.

field control are required and these motors are often employed in applications where the relationship between speed and torque is of a highly specialized nature.

Armature resistance control

The control of speed of a d.c. motor by means of inserting or reducing resistance into the armature circuit is most often met with in traction applications. In essence the control is actually effected by reduction of armature voltage, the difference between the applied voltage and that appearing at the motor terminals being absorbed by a variable resistance. This method of speed control is necessarily wasteful as the resistance dissipates a considerable amount of electrical energy when operating over a wide speed range. When operating with a large resistance in the circuit the speed of operation of the motor becomes very load-dependent. While in many cases this would be unacceptable, the feature can be effectively used as a simple form of torque control.

Choppers and pulse width modulation

Both choppers and pulse width modulation methods of control employ the principle of reduction of the effective voltage applied to the motor.

Choppers use static switching devices of the semiconductor type to interrupt the supply for short periods, thereby reducing the average voltage appearing at the motor according to the relative time the current is allowed to flow (see Fig. 6.3).

Pulse width modulation employs a similar switching technique with the variation that the on/off periods are modulated to provide a variable length to the on/off periods. Controllers of this type are usually met on applications requiring very fast response such as servo drives for machine tools. The production of variable average voltage from a pulse width modulation controller can be very smooth depending on the modulation frequency (see Fig. 6.4).

Ward Leonard-Ilgner control

The use of motor generator sets, i.e. a generator driven by some form of prime mover (for example, a steam engine, internal combustion engine, or constant speed electrical motor), provides a relatively simple method of obtaining a variable speed system

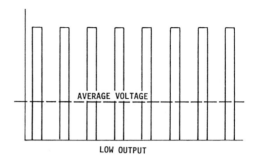

LOW OUTPUT

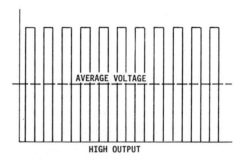

HIGH OUTPUT

Fig. 6.3   Output voltage derived from a d.c. chopper type speed controller.

from a d.c. motor. The various components of the system are as shown in Fig. 6.5.

Essentially the prime mover provides power to the system and the voltage delivered by the generator is varied by controlling the field. The speed of the d.c. motor is thus made to vary by means of armature voltage control. Such systems were in commercial use from the very earliest period of electrical history and indeed many similar systems are in use at the present time.

In certain applications the use of rotating generating systems can have inherent advantages; for example, in locations where no electrical supply is available from a distribution system some recourse must be made to the electrical generation of supplies at a local level. In these circumstances the design and choice of prime mover is usually dictated by the available fuels.

It should not be overlooked, however, that even with readily available electrical supplies the characteristics of motor generator sets can offer tangible advantages. The effective isolation of the electrical load from the main supply can offer benefits in situations

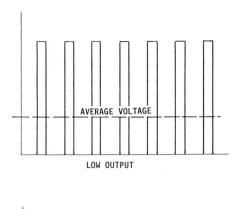

LOW OUTPUT

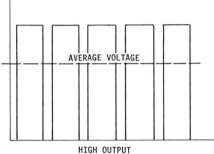

HIGH OUTPUT

Fig. 6.4 Output voltage derived from a pulse-width-modulated speed controller.

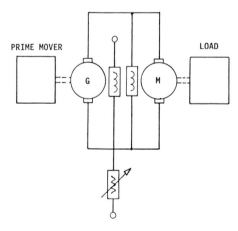

Fig. 6.5 Basic Ward Leonard-Ilgner system with constant speed prime mover driving a variable voltage d.c. generator feeding a variable speed d.c. motor.

Plate 6.1   Ward Leonard-Ilgner multi-generator set supplying power to a d.c. multi-motor drive system.

where there are severe restrictions on the potential interference that the installation is permitted to have on the primary supply. For example, in the rolling of metals very high peaks of load can occur as the material being processed enters the rolling mill. For such an application a generating set could be designed with a high natural inertia so that the energy stored in the mechanical mass can be allowed to flow into the load when required to relieve the electrical supply of the necessity of providing this peak of power and thus avoiding costly high transient demands on the power supply.

For such a generating set a considerable amount of information with regard to the process is necessary if the design is to be carried out conservatively. This information has to include data on power absorbed at all points in the duty cycle and this data is not always easily obtained, usually being supplied from empirical data derived from similar applications.

The rotating inertia of motor generator sets also gives them the ability to 'ride through' minor supply interruptions; this feature is particularly valid if the electrical supplies to the controlling switchgear motor and generator fields are derived from machines forming part of the main generating set, for under these circumstances any reduction in the armature voltage to the drive motor due to a small drop in speed of the generator will, to some extent, be offset by the reduction in flux in the driven motor caused by the corresponding fall in voltage from the excitation generator.

The relatively high thermal capacity of rotating plant permits economical design of the electrical machines which need only be rated for an average load taking into account the time cycle of the driven process.

The natural or basic inertia of the generating set can be artificially increased by the addition of a flywheel, thereby considerably increasing its ability to deliver energy to the load from the increased rotating mass. Sets of this type are often referred to as Ward Leonard-Ilgner (see Fig. 6.6). Their use is particularly evident in steel rolling mills and in mine winding hoists. A typical load cycle is shown in Fig. 6.7 and indicates the source of energy at various point in the duty cycle.

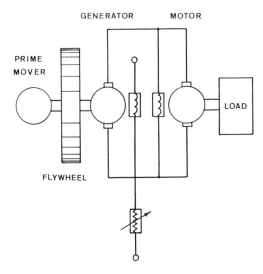

Fig. 6.6 Ward Leonard-Ilgner scheme with a flywheel mounted on the prime mover shaft, relieving the prime mover from the effect of peaks of load demand.

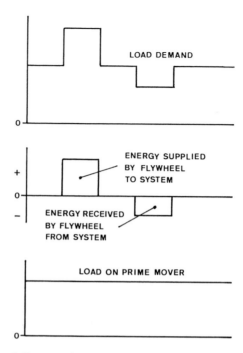

Fig. 6.7 Load diagram showing how peak demands of power are met by the flywheel in a Ward Leonard-Ilgner system.

Single-phase thyristor converters

Most industrial premises are provided with a three-phase supply, and as unbalance is considered undesirable, the use of single-phase converters is limited to those situations where the connected loads can be distributed with reasonable uniformity around the three phases or where the load is so small in comparison with total load as to be of negligible significance. In the majority of industrial situations the use of single-phase systems is usually limited to drives of up to 15/20 kW. The semiconductors comprising the converter can be connected in a half- or full-wave configuration. The bridge may comprise thyristors or a combination of thyristors and diodes. Examples of some typical circuits are given in Fig. 6.8, together with associated d.c. output waveforms. It should be noted that the bridge employing diodes in one half is limited in output range.

In normal practice a diode is usually connected across the output terminals of half-controlled bridges to provide a path for the current during the time when the converter is providing no output, thus avoiding excessive voltages being imposed on it and

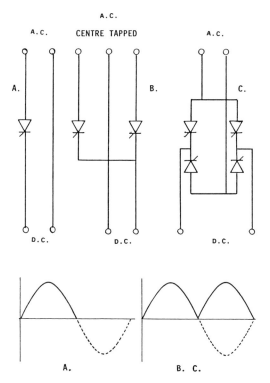

Fig. 6.8 Various methods of connecting rectifiers in a single-phase supply and the associated waveforms.

the motor by the sudden flux change. Single-phase converters are susceptible, when used on wide speed ranges, to an effect termed 'cogging', caused by the discontinuity of the current delivered to the motor at low speeds (see Fig. 6.9). This effect is particularly noticeable when such a drive is applied to a low inertia system where the momentary loss of torque due to the discontinuity of the voltage and current permits the system to fractionally decelerate, only to regain speed on restoration of the supply. The effect is to produce a pulsation of torque recognizable by a characteristic hammering or vibration.

Three-phase thyristor converters

The three-phase d.c. thyristor converter is perhaps the design most frequently employed for industrial drive purposes in power applications. The circuit configuration is shown in Fig. 6.10.

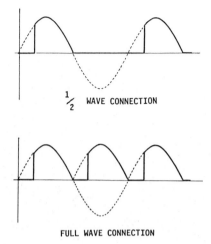

Fig. 6.9 Waveforms for half-wave and full-wave single-phase thyristor
controllers showing discontinuity.

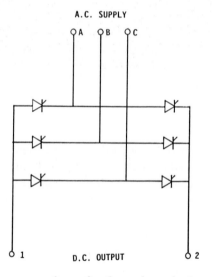

Fig. 6.10 Power connections of a three-phase thyristor controller.

The three-phase supply is connected at points A, B and C, and
the output is obtained at 1 and 2. The d.c. obtained from thyristor
bridges connected in this configuration provides an excellent
source of variable d.c. voltage and it is only at the extreme
bottom end of the range that the current can become discontinuous (see Figs 6.11 and 6.12) and risk of cogging occurs.

Plate 6.2 A small modular d.c. drive with control in forward and reverse motoring and re-generating (four-quadrant control).

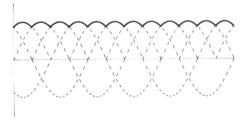

Fig. 6.11 D.C. output waveform from a three-phase full-wave (six-pulse) thyristor controller operating with a firing angle of approximately 60°.

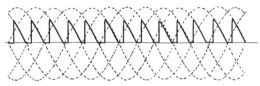

Fig. 6.12 D.C. output waveform from a three-phase full-wave (six-pulse) thyristor controller operating with a firing angle of approximately 140°.

Plate 6.3 Modular d.c. drive with the electronic control board removed to show the power connections.

Regenerative braking

The term regenerative braking is usually applied to braking of the controlled motor by means of rearranging its circuit configuration to allow the motor to act as a generator and return power to the supply. In fact the process of arranging for the motor to generate electrical power permits this power to be injected into any suitable load such as a resistance bank where the energy derived from slowing down the system can be dissipated in the form of heat. This is usually termed resistance braking.

To permit the motor to act as a generator it is necessary for the voltage generated by the motor to exceed that applied to its terminals and for the current to be reversed or alternatively for the excitation to be reversed so that its current can remain in the same direction with the torque reversed, i.e. braking instead of motoring.

Regenerative braking with thyristor supplies is therefore possible by means of field reversal as the direction of the current delivered from the bridge remains in the same forward sense. Control over the braking current is then effected by adjustment of the voltage delivered from the converter.

It is common practice for the field control to be achieved through a separate power supply which can be adjusted to obtain the braking characteristic required.

To achieve braking into a resistance load the procedure is simply to disconnect the motor from the d.c. supply and reconnect the motor terminals across the braking resistor. It is usual for either the excitation of the machine to be adjusted as described above to control the flow of current into the resistor or to arrange the resistor with several sections of reducing value, thereby maintaining the braking current and torque as the terminal voltage decays with speed.

Chapter 7
Harmonics

Most of the alternating current provided to consumers by the electrical supply authorities is derived from rotating generating systems which are designed to produce a waveform which is substantially sinusoidal, that is to say composed of a wave which is essentially of one frequency. The frequency chosen almost universally is either 50 or 60 Hz. In practice, however, the nature of the load supplied can distort this pure sine-wave and give rise to a complex wave. The commonest form of distorting load is the supply transformer, the distortion being caused by the nonlinear voltage/current relationship which occurs in a magnetic circuit, due to the saturation of the magnetic material.

Harmonics and drives

In more recent years thyristor drives have also become a major source of supply distortion. They create distortion because the current drawn from the supply is of an irregular non-sinusoidal form and in some instances discontinuous, the discontinuity being caused by the commutation of the thyristor device (Fig. 7.1).

Any symmetrical waveform can be shown by mathematical analysis to be composed of the addition of a number of pure sine-waves, of differing frequency or harmonics, added to a fundamental wave (Fig. 7.2).

The harmonic currents created by thyristor drives are related to the number of pulses which in a full-wave three-phase configuration is six and in a six-phase full-wave configuration twelve.

The harmonic currents present can be represented by the following formula:

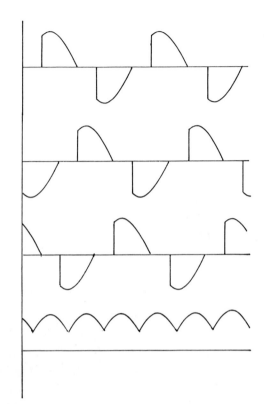

Fig. 7.1   Phase currents in a three-phase, full-wave, thyristor bridge operating with a firing angle of approximately 60° and feeding a resistive load showing the irregularity of the wave shape which gives rise to harmonic currents.

$$f_h = f_s \left[(K \times p) \pm 1 \right]$$

where: f_h = harmonic frequency generated
f_s = supply frequency
K = any whole number
p = number of pulses

Thus the theoretical harmonic frequencies created by a full-wave three-phase system on a 50 Hz supply can be shown to be:

$$f_1 = 50 \left[(1 \times 6) - 1 \right] = 250 \, \text{Hz}$$

i.e. 5 times the fundamental or 5th harmonic,

$$f_2 = 50 \left[(1 \times 6) + 1 \right] = 350 \, \text{Hz}$$

i.e. 7 times the fundamental or 7th harmonic,

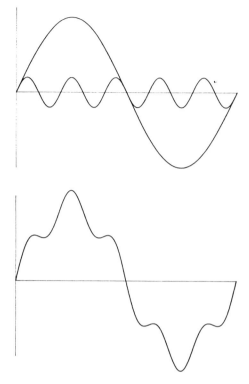

Fig. 7.2 Distortion of the fundamental sine-wave due to adding 20% of 3rd harmonic.

$$f_3 = 50 \ [(2 \times 6) - 1] = 550 \, \text{Hz}$$

i.e. 11 times the fundamental or 11th harmonic,

$$f_4 = 50 \ [(2 \times 6) + 1] = 650 \, \text{Hz}$$

i.e. 13 times the fundamental or 13th harmonic
    etc.

Similarly, for a six-phase system (12 pulse) 50 Hz.

$$f_1 = 50 \ [(1 \times 12) - 1] = \ \ 550 \, \text{Hz} = 11\text{th harmonic}$$
$$f_2 = 50 \ [(1 \times 12) + 1] = \ \ 650 \, \text{Hz} = 13\text{th harmonic}$$
$$f_3 = 50 \ [(2 \times 12) - 1] = 1150 \, \text{Hz} = 23\text{rd harmonic}$$
$$f_4 = 50 \ [(2 \times 12) + 1] = 1250 \, \text{Hz} = 25\text{th harmonic}$$
    etc.

The magnitude of these currents, for a full-wave system can be expressed theoretically by the relationship

$$I_h = \frac{I_L}{(k \times P) \pm 1}$$

where I_h = harmonic current
I_L = a.c. load current
k = any whole number
P = pulses

Thus for a full-wave three-phase 6-pulse thyristor drive drawing 100 A from the supply the values of the harmonic currents are theoretically

I_{h1} = 100 ÷ {(1 × 6) − 1} = 20 A of 5th harmonic
I_{h2} = 100 ÷ {(1 × 6) + 1} = 14.29 A of 7th harmonic
I_{h3} = 100 ÷ {(2 × 6) − 1} = 9.09 A of 11th harmonic
I_{h4} = 100 ÷ {(2 × 6) + 1} = 7.69 A of 13th harmonic
etc.

Similarly for a 12 pulse six-phase system:

1_{h1} = 100 ÷ (1 × 12) − 1 = 9.09 A of 11th harmonic
1_{h2} = 100 ÷ (1 × 12) + 1 = 7.69 A of 13th harmonic
1_{h3} = 100 ÷ (2 × 12) − 1 = 4.35 A of 23rd harmonic
1_{h4} = 100 ÷ (2 × 12) + 1 = 4.00 A of 24th harmonic
etc.

It should be emphasized that the above calculations produce only the theoretical values. The effect of supply reactance will reduce these levels and similarly a badly balanced thyristor bridge (i.e. a bridge where the firing pulses are not occurring at precisely the correct moment) could worsen the figures. Any investigation into the possible effects of a thyristor load on a supply should be undertaken with considerable care, taking into consideration the supply conditions as well as the nature of the load.

When it is intended to install substantial thyristor-controlled loads it is advisable to consult the supplier of electrical power to determine the compatibility of the proposed load with the supply system. Supply authorities usually set limits for the maximum load of this type that may be connected before the need for corrective action to reduce the potential harmonic content of the installation. There is a cumulative effect when additions to systems are made and this aspect must not be ignored when considering extensions to existing plant.

The presence of harmonic currents on supply lines is undesirable as they can interfere with electronic devices such as computers and can confuse the system of line signalling employed by the generating boards who may be using their transmission lines to send messages within the supply network; harmonic currents can give rise to breakdown of power factor correction capacitors and other equipment due to high voltages being created by electrical resonance; they can also excite torsional stresses on rotating electrical machines connected to the supply system. The undesirable effects of harmonics where these occur in sufficiently large quantities have sometimes to be reduced.

The initial approach is to increase the number of phases from three to six or even more in the case of very large systems. From the previous formulae it can be readily seen that by increasing the number of phases a substantial reduction in harmonic content can be achieved, thus permitting the connection of large loads onto a supply system. The popular way of increasing the number of phases from a given a.c. supply is by using specially constructed transformers with multiple secondary windings with the phases displaced relative to each other, or more simply by dividing the load and supplying each portion from a separate transformer, each with a mutual phase shift.

The effect of supplying harmonic currents gives rise to a distortion of the supply voltage and in order to ensure minimal disturbance to other equipment or users who may be connected to the same system, Engineering Recommendation G5/3 (a document of the now-defunct British Electricity Council) specifies that for an individual load the voltage distortion attributable to harmonic currents should not exceed, on a 415 V three-phase supply, 4% for the odd harmonics (3, 5, 7, etc.) and 2% for the even harmonics (2, 4, 6, etc.). For higher voltage supplies these limits are reduced to 3% and 1.75% respectively for 6.6 kV systems, and they are reduced further for even higher voltages.

Precise calculation of the effect of the harmonic content of a load on the supply requires a knowledge of the capacity and the regulation of the supplying system. For an individual harmonic load the distortion of the voltage can be calculated from the following formula:

$$\text{Voltage distortion (\%)} = \frac{P_c}{P_s} \times \frac{r}{100} \sqrt{\Sigma (H^2 \times I_h^2)}$$

where: P_s (kW) = supply capacity
         P_c (kW) = converter capacity
         r (%)   = system regulation
         H = harmonic number
         I_h = harmonic current

From the above it can be seen that the systems least likely to be adversely affected are those where the harmonic loads are small relative to the supply capacity and where the system has a low regulation, i.e. low source impedance.

It is of importance to note that the frequencies most likely to cause problems are the lower harmonic numbers and in particular those up to the 19th harmonic. Harmonic frequencies above this level tend to be heavily attenuated by the supply system and are therefore not so troublesome. If increasing the number of phases employed does not give an appropriate reduction in the offending frequencies resort has to be made to filters. These usually take the form of a tuned circuit or circuits designed to present a low impedance to the unwanted frequency and thus conduct the offending harmonic current away from the line. Wherever possible harmonic filters are avoided as their design is complex and the precise frequency to which they are tuned is usually a compromise. Additionally the diverted current has to be absorbed in the filter and thus filters are wasteful of energy and contribute to the loss of efficiency.

Engineering Recommendation G5/3, sets out the maximum recommended levels of harmonic current at various frequencies and supply voltages. These are shown in Table 7.1 together with the corresponding equivalent thyristor drive load that could be regarded as giving rise to the limiting values of harmonic current specified. Some caution must be exercised when consulting this table as it gives an indication of the expected levels only, and in designing a system proper calculations or measurements should be used to determine the position, particularly with regard to any existing loads on the supply system. The criteria for the connection of loads which could give rise to the generation of undesirable effects is the subject of ongoing debate among the various standards-making bodies, and the limits for the generation of low frequency harmonics will ultimately be included within the all-

Table 7.1 Recommended maximum harmonic current levels.

Supply voltage	*Harmonic current (A)*				*Typical maximum thyristor converter loads (kW)*	
	5th	*7th*	*11th*	*13th*	*6 pulse*	*12 pulse*
415	56	40	19	16	100	150
6600/11 000	10	8	7	6	800	1500
33 000	9	6	6	5	2400	3800
132 000	4	3	3	3	4700	7500

Source: ACE Report No. 15 and former Electricity Council Engineering
       Recommendation G5/3.

embracing standards relating to electromagnetic compatibility
(EMC) currently in the course of preparation in the European
Community.

Electromagnetic compatibility (EMC)

It has always been recognized that the operation of electrical
plant can give rise to an interaction with, or affect the performance of, other electrical equipment. The direct-on-line starting
of induction motors, for example, can draw large currents for
short periods and the effect of this on supply systems, where it
represents a substantial portion of the supply capacity, results in
a significant reduction in the system voltage due to the natural
regulation of the supply transformer and the associated cables
etc. In this situation it is imperative that other equipment connected
to this supply should be immune to this effect or at least able to
continue in satisfactory and safe operation.

The majority of the industrialized countries of the world have
over many years adopted various standards in order to regulate
the interactive effects of electrical equipment, although until
relatively recently these regulations concentrated chiefly on avoid-
ance of interference to radio reception. The increasing use of
computers and other sensitive electrical and electronic equipment
has given impetus to the necessity for widening the scope of
the regulations; the creation of the European Community has
emphasized a requirement that existing regulations be both
broadened and harmonized between participating countries.

The objective of the parties responsible for formulating the necessary regulations was to achieve a set of standards which would ensure that a user of electrical equipment could be reasonably assured of trouble-free operation and immunity from external events of an electrical nature which could interfere with, inhibit or bring about the malfunction of his equipment in the normal course of its utilization. Equally, it is necessary that the operation of such electrical equipment in one location does not generate or cause to be generated any electrical radiation or conducted emissions which could cause problems in another location. Quite apart from the nuisance value of interference with communications, the performance of electronic devices and computers, used to control the logic of industrial processes, may also be affected with potentially disastrous results.

EU directives

Manufacturers within the EU, where goods move from country to country relatively freely, may affix to their products a symbol, recognized throughout the region, which indicates a product's conformity with the necessary standards. This mark comprises the letters C and E in the form of the logo illustrated in Fig. 7.3. Predictably, the establishment of regulations acceptable to all participating countries took a long time, and since the first EU directive came into force in 1992 (Directive No. 89/336/EEC) there have been a number of amendments. No doubt there will be more amendments from time to time.

A period of time has been allowed for the transition from national standards to the new standards, and it is anticipated that the majority of participating countries will have completed legislation by 1996. It will then be mandatory for electrical equipment

Fig. 7.3 The European Common Market sign of conformity.

manufacturers to ensure that their equipment conforms to the new requirements. During the interim period, all apparatus purchased and put into service after 1 January 1992 must comply with any relevant national standards currently in force.

Directive No. 89/336/EEC covers virtually all electronic and electrical equipment except products for radio transmission which are governed by separate regulations. For the manufacturer and user of variable speed drives the regulations present a particular problem because this type of equipment can produce substantial radiation and transmission. Before considering the steps and methods that should be employed to achieve compliance it is appropriate to examine the wording of the EU directive, which provides the following definition at Article 4:

a) The electromagnetic disturbance it generates does not exceed a level allowing radio and telecommunications equipment and other apparatus to operate as intended.
b) The apparatus has an adequate level of intrinsic immunity to electromagnetic disturbance to enable it to operate as intended.

The wording of the directive is non-specific with regard to meeting actual standards, but is all-embracing insofar as it covers all types of electrical and electronic equipment operating in all situations. There are some specific exemptions from compliance with this directive but these are in the main covered by other directives which in themselves contain requirements relating to electromagnetic radiation, e.g. emissions from the ignition systems of motor vehicles. The directive is also not applicable to simple components intended to comprise part of a larger assembly, e.g. resistors or other small electronic items, and, at the other end of the scale, to very large fixed installations like power stations.

Standards

In order to ascertain compliance it is obviously necessary to have a set of limits or standards against which a piece of equipment can be measured. The creation of these standards was made the responsibility of CENELEC, the European electrical standards organization.

Of specific interest to the user or supplier of variable speed drives are the standards applicable to this sector of industry.

At the present moment these standards are in the process of finalization; on completion they will define the limits and test methods for drive systems and will be designed to ensure an adequate level of compatibility for all industrial and public locations.

Designing new equipment that complies with the requirements possibly presents a smaller challenge than modifying existing designs. When engineering and manufacturing with the regulations in mind at the outset each part or sub-assembly may be tackled in such a way as to ensure that the final product contains a high degree of inherent compliance, thus reducing any additional components that may be found necessary in the final assembly. Particular attention should be directed at assemblies and sub-assemblies operating at high frequencies, e.g. data transmission and processing circuits. The steps that can be taken cover a wide range, including careful routeing of connections, layout of printed circuits, reduction of the power levels at which control circuits are operating, careful attention to grounding and shielding. High frequencies are a particular problem as these can be radiated directly through the atmosphere and ultimately may require extensive screening.

Conducted and radiated emissions

Electrical drives, apart from the low power electronics circuits, also incorporate devices controlling comparatively large currents which are directly derived from the main supply. This gives rise to the propagation of interference which can be conducted into the supply circuits; this interference also has to be reduced to an acceptable level. Conducted emissions are usually dealt with by the application of filter networks whose function is to safely divert or contain the offending frequencies. The techniques adopted to reduce or eliminate radiated or conducted emissions and to reduce susceptibility are, in the main, shielding and filtering.

Shielding involves enclosing the equipment or item within a box or container that possesses the property of preventing emissions from escaping into the environment or from being imposed on the equipment from an external source. Such screens are made from materials that are conductive in varying degrees and it is important that the screen provides adequate protection in all directions. This requirement can often be at variance with

the need to provide cooling and in these cases recourse is made
to the installation of heat exchangers or to carefully designed
materials that are able to permit the movement of air while
providing the necessary electrical properties; these take the form
of perforated metals with the sizing of the holes and thickness
of the metal carefully graded. Screens and screened enclosures
should be soundly earthed, joints must be properly bonded and
terminations must be of an appropriate design. A wide range of
materials and terminating items are available for this purpose.

Filtering is a technique which is particularly applicable in
combatting emissions conducted from or to the equipment, and
these can take the form of both acceptance and rejection filters.
Typical filter networks are shown in Fig. 7.4. For the purpose of
protection broad-band filters are usually employed which possess
the property of reducing a wide range of interference. For an
equipment that is known to generate or be susceptible to a
specific range of frequencies the filters can be designed with a
narrow spectrum covering the specific frequencies it is desired to
attenuate.

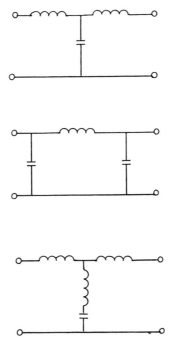

Fig. 7.4 Some typical filter networks used to remove or reduce the
effect of unwanted harmonic currents.

Installation

When installing electrical plant care should be taken that the installation procedures recommended by the supplier are strictly adhered to as all the efforts taken by a manufacturer to achieve compliance of his product can be set at nought by failure to adopt the correct techniques in this respect. As time progresses installing companies will become familiar with the correct techniques but responsible manufacturers will certainly, in the interim, be expected to provide sound instructions for the connection and installation of their plant.

The new legislation includes quite severe penalties for operating plant that gives rise to problems and includes provisions for the suspension of operation of such plant on receipt of a complaint. In this respect the ultimate user of electrical plant is advised to exercise caution when purchasing new equipment.

For the manufacturer of non-compliant equipment the penalties are equally severe: it may be banned from the EC and EFTA countries and the manufacturer or his representatives fined. Enforcement of the legislation will devolve on the authorities in the various member states, who will no doubt utilize the services of existing organizations, e.g. the Office of Trading Standards, who will be enabled to inspect and adjudicate, and investigate complaints concerning non-compliance.

Competent bodies

In order to achieve compliance with the requirements a manufacturer has to be able to demonstrate that he has tested his equipment in the appropriate manner or that he has constructed the plant in such a fashion as to satisfy an independent adjudicator that he has taken all the appropriate steps to ensure compliance. These adjudicators, who are appointed by the government, are known as competent bodies, and they will require satisfactory documented evidence of the measures that a manufacturer has taken in this respect. This evidence will take the form of a special file, the Technical Construction File (TCF), in which the manufacturer describes and identifies the equipment, detailing its design and the measures taken to ensure conformity. The competent body will examine this document and attach a report or issue a certificate confirming their findings, and this report will form part of the TCF.

The testing of equipment to comply with the regulations has generated considerable interest and in order to obtain meaningful information it has been necessary to construct special testing facilities to ensure that the equipment under test is effectively either isolated from its surroundings or situated in an environment for which the prevailing conditions are known and repeatable. The testing equipment is quite sophisticated and the data produced usually require interpretation. Testing of every piece of equipment manufactured is not necessary provided that a conforming type test has been carried out on each design. In this respect it should be particularly noted that even what at first sight may seem to be a minor change can result in an equipment failing to meet the requirements and it cannot be stressed too strongly that control over the manufacturing process must be strictly observed. The omission of a special gasket or failure to remove the paint from a surface designed to be conductive can result in leakage from a screening enclosure – and hence in non-compliance.

Chapter 8
Comparison of variable speed drives

Electrical variable speed drive systems can be broadly divided into a.c. and d.c. types, i.e. drives using a motor which is designed for use on a.c. or d.c. electrical supplies.

A.C. drives

Variable speed a.c. drives can be further subdivided into those types where the a.c. supplied to the motor is changed in frequency by some means external to the motor (e.g. the motor is supplied from an inverter delivering variable frequency) and those in which the speed variation is achieved internally by special windings sometimes in conjunction with external regulating devices.

The basic features of various types of a.c. drive are briefly summarized in Table 8.1.

It should be noted that the ranges indicated in Table 8.1 are typical of industrial applications and they do not imply that any

Table 8.1 Salient features of a.c drive systems.

Variable frequency a.c. drives	Speed range	Power range (kW)
Pulse width modulation	100 : 1	0–5000
Quasi-square wave	50 : 1	0–10 000
Cyclo-converter	100 : 1	500–10 000
Slip recovery (Kramer)	2 : 1	100–5000
Variable speed a.c. motors		
Commutator motor (Series control)	5 : 1	0–10
Pole-changing motor	2 : 1	0–50
Commutator motor (Schrage N–S)	10 : 1	50–3000
Slip recovery (Kramer)	2 : 1	100–5000

particular drive type is technically limited to these ranges. Slip recovery drives are included in both sections because this form of variable speed system can be designed as an integrated motor, the slip energy being recovered with a rotating machine coupled to the rotor or by static inverter recovery, as explained in Chapter 2.

D.C. drives

It is convenient, for purposes of comparison, to divide drives into static and rotary types (see Table 8.2).

Static

Static d.c. converters can be subdivided into single- and multi-phase types.

Single phase
Single-phase converters are in general only used for smaller drives, up to about 15 kW, due to the poor waveform and the problems that can arise from this with commutation of the motor current. Where single-phase drives are used it is quite common practice to place an inductance in series with the load circuit to ease commutation by smoothing the waveform.

Multi-phase
Multi-phase d.c. converters are used extensively in all sizes; exceptionally wide speed ranges can be achieved by using this type of converter, 100 : 1 being entirely typical with constant torque.

Table 8.2   Comparison of d.c. drive systems.

	Efficiency %	*Power factor*	*Power range*
Rotary converter	80	0.85/0.95*	not limited
Static single-phase	98	0.1/0.83**	up to 15 kW
Static multi-phase	98	0.1/0.8**	not limited

 * Power factor is dependent upon the type of prime mover.
** Power factor is dependent upon operating speed range, and the control
   margins allowed in the design.

Rotary

The efficiency of static converters is in general considerably higher than rotary converters but rotary conversion systems have the advantage of being able to operate at high power factors over their whole operating range whereas the power factor of static converters can fall to a very low figure at the bottom end of a wide speed range. Rotary converters are also able to generate very smooth d.c. to the drive motor over the entire speed range, and this makes them particularly suitable if the load is ultra-sensitive to the torque pulsation which can arise with static converters.

From Table 8.2 it can be seen that the method of comparison is not the same in both the d.c. and a.c. cases. All comparisons are complicated by the criteria used to make the observations and these are not easily transferred from a.c. to d.c. drives due to the fundamental differences in their control and configuration. As an example of the problem, if it was decided to compare a.c. and d.c. drives on the basis of first cost only, and the application was of a very simple type requiring only operation at two different speeds, the obvious choice would be a pole-changing motor. If, however, the requirement was for smooth operation over a range of speeds down to, say, half speed, some other drive would have to be selected because pole-changing is not technically capable of performing this duty.

Some other method of comparison is therefore desirable that puts into context the performance required from the drive system. This is the reason for selecting the parameters included in the tables above, which, while different, highlight the main features. To summarize, the simpler the application the more elementary the drive can be. Simple applications can be defined as those requiring only a limited speed range, low starting torque, constant or falling load torque with reducing speed and with no limitations on the generation of harmonics or operation at low power factors.

Chapter 9
Selection of variable speed drives

The selection of a drive for any particular duty needs to be approached with a number of considerations in mind; attention should be given at each stage to the requirements of the application. For some applications, the choice is probably, not definable in precise terms and a number of drive types, both a.c. and d.c., could be employed satisfactorily. In such cases the user may well select the type of drive system with which he feels most comfortable. Such decisions are usually made on the basis of familiarity with a particular type of drive or one that the user feels most able to maintain adequately.

For most practical purposes the selection can be made from an examination of:

- The load characteristics
- Environmental issues
- Maintenance
- Cost and running cost.

Each of these issues should be examined in detail and in many cases the choice of drive will become obvious.

Load characteristics

The type of load to be driven has a direct bearing on the selection of a drive type. It is necessary for the user to be aware of the speed and torque characteristics demanded by the driven load in addition to the actual power requirements for the application for which the drive is to be selected. For the best selection the drive system should be as accurately matched to the load requirements as is possible.

The nature of a driven load can be defined in a number of significant ways, typically:

- Torque versus speed
- Impulse loads
- Power requirements
- Coupled inertia/starting torque
- Torsional rigidity and mechanical resonance
- Speed range
- Control response.

Torque versus speed characteristic of a load

The majority of applications can be classified in groups depending upon how the torque and power requirements vary with speed of operation.

Centrifugal loads are usually met with in duties involving the movement of liquids or gases, i.e. pumping or fan applications. The drive has to be designed in these applications to meet the maximum demand for power from the driven pump or fan, which occurs when the machine is operating at its maximum output or delivery.

To achieve precisely the desired flow or pressure where a constant speed drive is employed it is common practice to use a regulating device such as a valve or some other restriction to flow. This is obviously inefficient as energy is absorbed in the valve and in the turbulence which is often created. A more energy-efficient concept is to vary the speed of the pump or fan in accordance with the process requirements. To achieve this satisfactorily it is of course necessary to be aware of the characteristics of the load in respect of the way in which delivery or pressure varies with speed. In the case of the majority of centrifugal loads this is not a linear relationship but follows a 'characteristic curve' which is unique to the design of the pump or fan.

Pumps

Pumps can be broadly divided into two groups, positive displacement and centrifugal types.

Positive displacement pumps are usually designed to deliver a specific quantity of the fluid being moved for each movement of the pump piston or rotor and can be either rotary or reciprocating in design. When operating with normal delivery, the load is essentially a constant torque type, but if the output becomes restricted or blocked or the pump is being employed to increase

Plate 9.1 Variable speed totally enclosed a.c. motor with top-mounted heat exchanger. (Courtesy of Laurence, Scott and Electromotors Ltd.)

the pressure in a closed vessel the load torque can rise dramatically. Positive displacement fluid pumps are sometimes, therefore, designed to operate under stalled conditions and this factor must be considered in the drive rating and selection. In such cases it is usual to either rate the drive for full torque at stand-still or provide for some rapid-acting overload device to safely stop the drive under stall conditions.

Similarly for positive displacement air compressors the drive has to be capable of providing sufficient torque to get the pressure vessel up to the required working pressure. Both considerations are in addition to the selection of the drive based solely on the power required for normal running operation and delivery. For these applications it should also be borne in mind, particularly for fluid pumps, that the starting torque can be high if the pump is required to start with the system pipework full of fluid because under these conditions the entire system has to be moved from rest to its operating condition and both static friction and the operating head have to be overcome. The chief considerations in selecting a drive for these types of load is, therefore, one of correct selection of torque and overload capability.

While positive displacement pumps are used in large numbers

in industry for specific applications, the centrifugal pump is used for more general applications. In this type of pump an impeller rotates continuously to accelerate the medium being moved or transferred. The centrifugal pump behaves in a very different manner from the positive displacement pump in that the shaped impeller (the rotating part of the pump) is designed in such a way that the faster it is driven (within certain physical limitations) the greater the pressure and volume delivered. This characteristic gives it a quite different performance curve of speed versus power which is non-linear and in fact closely resembles a cubic law, i.e. the power required/absorbed is proportional to the cube of the speed. A typical load/head (pressure)/speed curve is shown in Fig. 9.1.

It can be seen that to reduce the flow from a pump operating at constant speed by applying some restriction to the output or flow the head (pressure) increases and conversely the delivery decreases. As observed earlier, to achieve the desired flow rate by this means is wasteful of energy and considerable power savings can be achieved by changing the speed of the pump. As the characteristic curve of the pump is of a cubic law type the changes in speed do not have to be large to effect a considerable change in delivery.

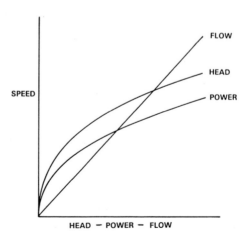

Fig. 9.1 Typical head/flow/power characteristic curve for an hydraulic centrifugal pump.

In actual practice the amount of energy that can be saved is also influenced by the variation in flow required and the period for which the pump has to operate at reduced speed – and also the amount of static load that exists in the pipeline system. In general the greater the static head the less the savings in energy that can be achieved. The reason for this is that the effect of static head on a pump is to change its characteristic curve.

The curve shown in Fig. 9.2 shows the effect of changing the static head on the power requirements and hence the ability to save power by variable speed operation. The torque required for pumping applications varies over the speed range with maximum torque only being required at top speed; starting torques are in general low.

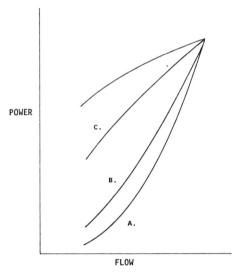

A.   -   NO STATIC HEAD

B.   -   25% STATIC HEAD

C.   -   30% STATIC HEAD

Fig. 9.2 Effect on power requirements of a centrifugal pump of changing the static head.

Fans

These operate on similar principles to centrifugal pumps and exhibit similar characteristics. The power demand and speed curves differ in some degree, however, due to the nature of medium that is being moved. As with liquid pumps the output from a constant speed fan can be modified by the use of dampeners or baffles or by adjusting the inlet conditions. It is obvious, therefore, that similar energy savings, by using variable speed, can be achieved with fans to those obtainable for applications involving fluid pumps. Typical curves of pressure versus flow are shown in Fig. 9.3 and the selection of a drive for fan load applications can be carried out in the same way as for centrifugal pumps.

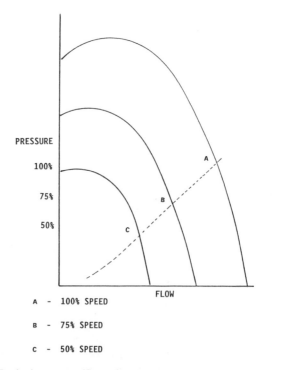

Fig. 9.3 Typical pressure/flow characteristic curve for a fan.

Conveyors

Conveyors comprising belts or buckets are used extensively for the transportation of bulk materials and for the movement of goods in warehouses. The load is usually characterized by the requirement for a relatively high starting torque compared with its running load and in some cases the conveyor has to be able to operate with a reversal of torque (overhauling loads); the drive may also have to be able to provide electrical braking to stop or slow down the conveyor. The power demanded by a belt conveyor is proportional to the velocity and tension in the belt. The calculation of belt tension is obtained from a knowledge of the effective torque (F_b) at the driving pulley. This figure is arrived at by the summation of the forces necessary to move the empty conveyor belt (F_e), the force required to move the load along the conveyor (F_l) and that required to raise (or lower) the load (F_r) if the conveyor is not horizontal. Thus:

$$F_b = F_l + F_r + F_e \text{ (see Fig. 9.4)}$$

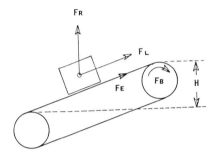

FR = FORCE TO RAISE LOAD TO HEIGHT (H)

FB = EFFECTIVE TORQUE AT DRIVING PULLEY

FE = FORCE TO MOVE EMPTY CONVEYOR

FL = FORCE TO MOVE LOADED CONVEYOR

Fig. 9.4 Diagram of the forces acting on an inclined conveyor.

Extruders

In many ways extruders can be regarded as similar to positive displacement pumps in that the materials to be extruded are conveyed or delivered to the extruding die by means of a screw-type conveyor which is essentially a form of positive displacement pump. Particular care must be taken when selecting drives for extruders as these are frequently required to start against a very high resistance and on occasions operate under stalled conditions for short periods. The selection of a suitable motor power and type is usually done by the extruder manufacturer who gives consideration to all aspects of the load including the viscosity of the extruded material and the requirements of the process for high extrusion pressures. Much of the data necessary for selection of extruder drive powers are obtained empirically from observed site data.

Winders and lifts

Both winders and lifts for mining and passenger purposes are the subject of very strictly regulated standards and the selection of drives for these duties is usually based on considerations of the nature of the loads to be carried. Passenger lifts have a requirement for very gentle starting and stopping and for accurate levelling. The ability to start with full load is inherent in the rating requirements.

Coilers and uncoilers

Coiling or uncoiling of materials is often carried out in continuous process industries as it provides a compact method of transportation of long lengths of material. Typical coiling processes are used in the sheet metal and paper-manufacturing industries. As most continuous processes are designed to run at an optimum speed for the product under manufacture it follows that the coiling

Plate 9.2 D.C. motor with integral safety brake for lift duty. (Courtesy of Bull Electric Ltd.)

system must run at a progressively slower speed as the diameter of the coil increases – and, conversely, at an increasing speed if the material is being uncoiled – if the tension is to remain constant. To achieve this, coils are often driven by friction through a roll which is pressed into contact with a spool around which the material being coiled is wound. The driven roll is then required only to run at a constant speed which corresponds to the linear speed of the material being processed.

The driving motor with this arrangement operates at a constant speed (the line speed) and delivers a constant torque which equates to the tension in the material plus the operating losses. However, in many cases it is considered undesirable to place drive rolls in direct contact with the reel in order to avoid mechanical damage to the material being processed. Under these conditions the drive system is normally connected to the spool at the centre of the coil and is required to operate to provide not only a changing speed but also a changing torque. The operating characteristic of such a drive is shown in Fig. 9.5. The selection of a drive system for such applications is usually dictated by the type of drive that can produce such a characteristic and these are most often d.c. or vector control a.c systems. In selecting a suitable drive for coiling duties it is also necessary to determine the accelerating and decelerating loads together with the steady running load so that an appropriately sized system can be employed.

A typical calculation for the drive requirements for a sheet-processing and reeling machine (see Fig. 9.6) follows.

Process machine data

Maximum linear line speed	200 m/min
Width of material	2 m
Tension in material	8 kg/cm
Accelerating time	10 s
Decelerating time	10 s
Reel and core weight	25 000 kg
Mechanical losses at reel	5 kW
Power used in process section	50 kW
Maximum diameter of reel of material	1.5 m
Diameter of core	0.3 m
Inertia of process section referred to motor shaft	$10 \, kg/m \, s^2$
Speed of motor of process section corresponding to maximum line speed	1000 rpm

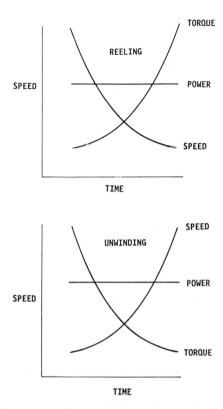

Fig. 9.5 Coiling and uncoiling duty power, speed and torque curves.

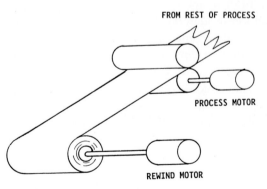

Fig. 9.6 Typical arrangement of the coiling or rewinding section of a processing machine.

The rewind section is considered first.

Power required to produce required tension in material when running at maximum speed

Total tension (T) = 8 × 200 = 1600 kgf = 15686.27 newtons

Power is proportional to n × t

where: n = speed (radians/s)
           t = torque = T × radius of reel (r)

and as: speed (n) = 2π × (L ÷ C)

where: L = linear line speed (m/s)
           C = circumference of reel [2π × radius (r)]

therefore: Power α 2π × (L ÷ 2πr) × t × r

rewriting and cancelling: Power α L × T

thus: kW = (L × T) ÷ 1000 (1 kW = 1000 newton metres secs)

For the drive under consideration this gives:

Drive power = [(200 ÷ 60) × 15686.27] ÷ 1000 = 52.29 kW

Power required to decelerate the reel in 10 s

Weight of reel and core = 25 000 kg

Radius of gyration (assume a solid cylinder)

$$k = \text{radius} \div \sqrt{2}$$
$$k = 0.75 \div \sqrt{2} = 0.531 \text{ m}$$

$$\text{Inertia} = I = (W \times k^2) \div g$$

where: W = weight (mass)
           k = radius of gyration
           g = 9.81

For the section under consideration this gives:

$$I = (25\,000 \times 0.531^2) \div 9.81 = 718.56 \text{ kg/m s}^2$$

Total energy possessed by full reel and core when operating at maximum line speed:

$$E = \tfrac{1}{2} \times I \times \Theta^2$$

where: I = inertia

Θ = angular velocity (radians/s)

The angular velocity for a 1.5 metre diameter, reel rotating to give a linear speed of 200 mpm (3.33 mps) is:

$$3.33 \div (\pi \times 1.5) = 0.7072 \text{ rps or } 4.44 \text{ radians/s}$$

therefore $E = \tfrac{1}{2} \times 718.56 \times 4.44^2 = 7082 \text{ kg/m s}$

As 1 Newton metre $= 0.102 \text{ kg/m}$

$$E = 7082 \times 0.102 = 69438.2 \text{ Nm s}$$

Thus the average power to decelerate the reel in 10 s is:

$$\text{Power (watts)} = 69\,438.2 \div 10 = 6943.82 \text{ W}$$
$$\text{or } 6.94382 \text{ kW}$$

This is the average power required. If the deceleration is assumed to be to zero speed at a linear rate, the peak value of power required will be twice this figure, i.e. 13.88 kW.

Therefore for this section:

(1) Normal running power $= 52.29$ kW
(2) Power required during deceleration $= 52.29 - 5 - 13.88$
$= 33.41$ kW

It should be noted that the requirement to accelerate this reel in 10 s would give the following power demand:

$$52.29 + 5 + 13.88 = 71.17 \text{ kW}$$

Normally a motor would be selected to provide the tension power and the losses, i.e. $52.29 + 5 = 57.29$ kW with an overload capability of approximately 25% for 10 s.

Considering now the process section, if the tension in the material entering the process is equal to the tension on leaving, for the purposes of this calculation it can be ignored.

Inertia $= 10 \, kgm \, s^2$

Motor speed $= 1000 \, rpm = 16.66 \, rps = 104.67$ radians/s

Energy stored in the section at maximum line speed:

$$E = \tfrac{1}{2} \times 10 \times 104.67^2$$
$$= 54779 \, kg/m \, s$$
$$= 537049 \, Nm \, s$$

Thus the average power required to accelerate the section in 10 s is:

$$Power \, (W) = 537049 \div 10 = 53704.9 \, W$$
$$or \; 53.70 \, kW$$

The peak power (assuming linear acceleration from zero) will be twice this figure, i.e. 107.4 kW.

The total power required for this section during acceleration will therefore be:

Accelerating power + running power

$$= 107.4 + 50 = 157.4 \, kW$$

and during deceleration

$$= 107.4 - 50 = 57.4 \, kW$$

A suitable motor for this section would be rated at approximately 80 kW with an overload capability of twice full load for 10 s.

General applications cannot be dealt with adequately here as they are so varied and numerous, but in all cases it can be said that the basic rules of selection given for the above typical applications should be applied as appropriate and the drive as closely matched to the load as possible.

The basic data required can be summarized as follows:

- Starting torque
- Torque/speed characteristic
- Braking
- Steady continuous running load.

Wherever possible all the above characteristics should be given consideration and in some cases on examination of these requirements the drive selection can be made directly. Table 9.1 shows some typical applications divided into their suitability for a.c. or d.c. drives.

Impulse loads

This type of load is in some respects one of the more difficult for electrical drives and a precise specification is important if a correctly sized system is to be selected. The mathematics of the analysis can be quite tedious as it is necessary to determine the magnitude and direction of the flow of energy throughout the duty cycle. With an impulse load a considerable amount of energy can flow from the rotating masses and accurately sizing a drive for this duty requires a good knowledge of mechanical engineering and, in particular, momentum and energy.

Problems can also be experienced with impulse loads where the repeated impulse is transmitted to the driving motor through the mechanical coupling, causing deterioration of the motor. This is usually overcome by using a coupling with some energy-absorbing characteristic like a spring or rubber buffer to act as an isolating device. Where impulse loads are encountered it is not at all unusual to find that the system designer has added inertia to the rotating masses to assist in providing energy to meet the pulsating demands of the load. Such inertia can be provided by using a flywheel mounted between the motor and the load and in some cases, by the simple selection of a larger motor (greater rotor inertia), the same effect can be achieved and the flow of electrical

Table 9.1 Drive suitability for application.

Duty	a.c.	d.c.
Pumps	*	—
Fans	*	—
Extruders to 100 kW	—	*
Extruders above 100 kW	*	*
Conveyors to 5 kW	*	—
Conveyors above 5 kW	*	*
Winders/lifts – goods	*	—
Winders/lifts – passengers	*	*
Coiling, small	*	*
Coiling, large	*	*

power smoothed. It is desirable that the added inertia be incorporated between the motor and the load so that any energy delivered by the flywheel is supplied directly to the load and does not have to pass through the motor shaft.

Power requirements

The actual power requirements of the drive system at a steady operating speed is in the majority of cases less than the peak rating of the selected drive motor. The actual power selected must be sufficiently large to accelerate the system and provide in addition a degree of overload insofar as circumstances permit.

The power to run a machine with a constant torque characteristic can be determined by measurement of the torque required to keep the shaft turning at a constant rate. The product of this torque and the maximum speed provides the rating of the drive.

$$P = n \times t \times k$$

where: P = power in watts
         n = speed in rpm
         t = torque in kg m
         k = constant of conversion = 1.022

For torque in lb ft
         k = 0.1421

Coupled inertia and starting torque

These two factors are sometimes combined as they are often considered to be directly related. This is only true, however, if the effects of static friction are ignored. To break away a load from rest a finite torque is required; this is quite different from the load required to accelerate the masses of the driven machine. However, in selecting a drive the two effects combine to create a torque demand in excess of the normal steady-running load. Static friction is very difficult to predict and it varies enormously depending upon the driven machine. It is usual to add a notional sum for this requirement. This is generally associated with the application and is usually derived from static torque tests using a torque arm and spring balance or some other simple mechanical means of assessment. If it is anticipated that the static friction could be high, measurement should be taken or advice from the machine manufacturer sought.

To achieve a steady running speed from rest the mass of the driven machine must be accelerated and the additional energy which is then stored in the rotating mass must come from the drive system. This energy is additional to that absorbed by the machine in doing work under steady running. The accelerating power is defined as the rate at which energy has to be supplied to the system to achieve a steady-state operating speed. Obviously if the system is to be accelerated at a high rate more power will be required than if acceleration takes place slowly. However, the total energy supplied and stored is the same in each case to achieve the same steady-state operating speed. From this it can be seen that the accelerating rate and the coupled inertia has a direct bearing on the selection of the motor/drive rating. As most drive systems are rated on power, usually specified in kW at maximum speed, care must be taken in sizing a drive system to ensure that the selected converter/inverter and motor combination will provide the requisite power and torque at all conditions of operation.

For a drive system that also has to provide electrical braking the same calculations must be made in respect of the deceleration of the system. Under deceleration conditions it should not be overlooked that energy absorbed by the load is assisting braking, as is the energy which is supplied to the electrical losses of the system. This in general means that if the accelerating and decelerating times are similar the drive system need only be sized for the accelerating conditions. A wide divergence in the accelerating and decelerating times would indicate the need to calculate both conditions to establish correctly rated equipment.

Torsional rigidity and mechanical resonance

All mechanical systems which transmit power have a certain degree of elasticity or flexibility. This feature can be regarded as a spring and as such has the ability to both store and deliver energy at a certain rate. The 'stiffness' of a system and the mass to which it is connected determines the rate at which it will oscillate when subjected to a disturbing impulse. The frequency of this oscillation is called the 'natural frequency' of the system. This can vary over a very wide range of values and for complex systems with several masses connected by flexible shafts there may be more than one natural frequency. If a system is subjected to an imposed disturbance at the natural frequency a resonant condition will occur which, with inadequate damping, could result in violent oscillation of increasing magnitude and ultimately severe damage. In the selection of drives it is therefore important to avoid utilizing a type that could impose a pulse of torque at or even near to the natural frequency of the mechanical system.

Speed range

The speed range over which a system is to operate is a critical factor in determining the type of drive to be employed. Many applications can be satisfied with speed ranges as small as 2 : 1 or 3 : 1, and for these situations the type of drive employed is not limited. However, once the speed range exceeds 10 : 1 it becomes necessary to consider in more detail the performance of the drive and the need for stable operation at lower speeds. Speed range cannot be considered in isolation for, as indicated earlier, the torque requirements at lower speeds are an important consideration.

If we consider the commonly met situation of constant torque over the speed range this can be easily met by d.c. systems with multiphase operation. Some a.c. drives operating over a wide speed range can give pulsations of torque or even a substantially smaller torque at low speeds compared with that obtainable at the maximum speed and it is for this reason that it is often necessary to derate or separately ventilate an a.c. drive which is required to operate over a wide speed range, giving rise to inefficient operation. Examples of the speed and torque characteristics of d.c. motors and a.c. motors with variable speed control are shown in Fig. 9.7.

Control response

The ability of a drive system to respond to a command to change speed or to meet a sudden demand of power can be an important factor in the selection of a drive. Speed response is determined by the rate at which energy can be provided or removed from the driven system. In situations requiring a very rapid reaction every effort is made to reduce the inertia of the drive itself, which can represent a significant part of the total system inertia. This can be effected by utilizing motors of special design which have their rotors constructed with reduced mass.

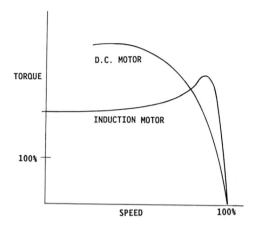

Fig. 9.7 Torque/speed characteristic of d.c. and a.c. induction motors.

A further technique is to reduce the inductance of the circuit which permits the rapid increase of current and consequently the torque delivered to the system. Many examples of highly responsive drives can be seen in industry where servo-mechanisms and some machine tools utilize disc type, permanent magnet motors or motors constructed with rotors which are long relative to their diameters. Stepper motors, which are fed with pulses, are also used for rapid-acting drives in low power applications.

Environment

In a considerable number of industrial applications the environment in which the drive is to be housed is of a reasonable industrial standard (i.e. dry and clean), and in these locations no particular preference can be expressed for any form of drive or system. However, in some instances where a drive or more particularly a drive motor is exposed to harsh or hazardous conditions, the drive selection can be quite critical. In general, a.c. drives employing simple induction motors which can be designed for enclosed operation are to be preferred for such applications. Drives using d.c. motors can be used in these situations if the torque or load characteristics render them more suitable than a.c. systems. It is, however, more difficult and expensive to manufacture such motors and it is often more appropriate to house a standard d.c. motor in a separate environment with the driving shaft protruding through the wall of an environmentally protected enclosure.

In some cases where potentially explosive or flammable gases are present the need to purge the motor enclosure with fresh air or nitrogen in order to disperse any hazardous gas is dictated prior to start-up. In such locations a variety of electrical machines can also be procured which have received explosion-proof certification. These motors are built into a special enclosure which effectively prevents ignition of the surrounding atmosphere by sparks originating within the enclosure.

Maintenance

All electrical drive systems require some maintenance, although it is true to say that modern electrical and electronic equipment is becoming very reliable, particularly if the electronic circuits have been designed and constructed to top industrial quality with all the manufacturing control this implies. When selecting a drive some consideration should be given to the maintenance required both for the converter or inverter and the motor.

Alternating current motors require only occasional cleaning and of course regular attention to lubrication. Direct current motors on the other hand require regular attention and examination of the brush gear and commutators; if this is carried out on a regular schedule the life and performance of a d.c. motor can be equal to that of an a.c. machine of similar size. Thus for a site with only limited access and a difficult environment an a.c. drive could be seen as the logical choice. It is equally true, however, that a.c. drive controllers are more complex and contain more components than d.c. controllers and where the level of maintenance is limited technically the less complicated and more easily understood d.c. drive might be seen to be preferable.

Cost and running cost

The actual running cost of a variable speed drive system is a function of the efficiency and the load being absorbed by the application. The typical efficiency of a.c. motors of the induction type used in variable speed drive applications is in the range 80%–85% for powers between 0 and 10 kW and 88%–95% for powers between 20 and 300 kW. These efficiencies are stated at maximum speed and a typical efficiency curve is shown for variable speed applications in Fig. 9.8. The inverter feeding such a variable speed motor would have an efficiency of about 96%. Thus the maximum overall efficiency of the system will be in the order of 80%–90%. Motors of the d.c. type are typically between 79% and 85% efficient for 0–10 kW and between 88% and 94% efficient for powers between 20 and 300 kW. It will therefore be seen that there is no significant power saving between a.c. and d.c. motors. The d.c. controller, however, has a typical efficiency of 98% which is somewhat better than an a.c. inverter.

The cost of initial equipment varies with the rating of the drive system and in general the larger the drive the smaller the cost expressed in terms of cost per kW. At the present moment the cost of d.c. drive systems is less than a.c. systems of similar size, rating and performance, and this is particularly true in the medium power rating of 100–500 kW. This cost comparison takes into account the fact that d.c. motors of equivalent rating to a.c. motors are more expensive and a.c. systems are more costly because of the more complex power and electronic circuits required to achieve the performance. In considering first cost it is necessary also to ensure that the actual rating assigned to the selected drive takes into account starting torque and low speed torque requirements and that drives are not simply compared on their nominal top speed ratings.

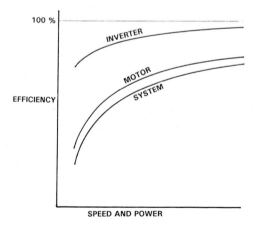

Fig. 9.8 Efficiency of the elements of an inverter drive system.

To determine running costs truly it is important that consideration be given not only to the nominal efficiency but also the duty cycle and maintenance costs.

To determine the true operating cost of a variable speed system it is necessary to calculate the power demanded over the operating cycle. A typical calculation for energy consumption is shown in Table 9.2. It should be noted that the above calculation has been conducted on the basis of one hour of operation. In a normal industrial situation this period would be adjusted to cover a complete operating cycle that may extend over a considerably longer period of time.

Individual drives

In industry the most frequently met variable speed drives are those which are required to perform a function quite independently from any other drive, by requiring the drive to stop and start, perhaps in some special sequence and then run at a speed determined by the process or an operator. This situation is relatively straightforward and is often termed a 'stand alone' drive application. Drives for such applications can be purchased as completely engineered systems which can be directly applied to the duty for which they are required and 'commissioned' from handbooks or instruction sets supplied by the manufacturer. Many such packaged systems, both a.c. and d.c., are available and it is of interest to note that in this particular field a.c. drives are becoming predominant, particularly in the low and medium power ranges up to about 100–150 kW.

Table 9.2 Pump application – energy requirements.

Delivery (%)	Power (kW)	Time spent at each speed (%)	kWh for 1 hour of operation
100	100	20	20.0
80	64	30	19.2
70	49	40	19.6
50	25	10	2.5
		Total energy	61.3

Grouped or sectional drives

A different category of drive system is that in which a number of drives are required to operate as an integral system as part of a process that has sections in which the speed of each drive is critical relative to the others. Most continuous processes that have a variable speed requirement with a number of driven sections fall into this category. Typical of these types of processes are the steel, papermaking and processing industries. In such systems not only the overall speed of the process has to be controlled and adjustable but also the speed 'section-to-section' of the process.

In the demand for increased output, process machines are required to operate at faster and faster speeds and this calls for a greater emphasis on design for accuracy of control. Modern sectional electric drives employ digital systems for control of intersection speed and this provides a highly stable and repeatable form of control. To achieve an integrated performance from a multi-motor system considerable care has to be exercised on the selection of the power of the various section drives. It is necessary to ensure that each drive can not only meet the power demanded for acceleration and steady running of its own section but also match the control response of the other units comprising the system.

This aspect of the design requires a good knowledge of the standing load and inertia of the complete system and any possible peak loads that might be encountered. From this data a calculation of the drive requirements can be made. If the system is section-speed-dependent, a useful check can be made by determining the relationship of inertia referred to the motor shaft and the power rating of the drive and motor. This gives an indication of the ability of a drive system to respond to a change in speed demand and is sometimes referred to as the 'accelerative response' of the drive. If there is an unduly large difference in this ratio within the various sections of the machine a further check should be made to ascertain the minimum acceptable response to a change in speed demand. Widely differing response time can give rise to stresses in the processed material as the speed between sections can vary while the system reaches its new stable state of operation after a disturbance.

The above must not be confused with the ramp times which are built into the drive for the purposes of limiting the rate of normal

acceleration. It is not possible to give a set of performance criteria
for this function as it is a factor related to the strength of the
processed material and the effect small tension variations might
have on it. Obviously it is of less importance with a material such
as steel, which has a high tensile strength and is usually processed
at fairly high tensions, than in the case of very thin or weak
materials or where the process variables are altered by small
fluctuations in speed and tension and which could give rise to the
production of unacceptable material.

Grouped drive systems usually employ an overall master reference signal by which all the sections are controlled. Adjustment
of this signal causes all the motors to increase or decrease speed
together. Individual speed trimming can be carried out to adjust
the precise ratio of speed between each section (see Fig. 9.9).
A variation of this form of control is a cascaded system wherein
each section drive receives its speed demand signal from the
preceding section. In this form of control the cascade can be
arranged with each succeeding motor receiving its speed signal
from a speed measuring device in the preceding section (see
Fig. 9.10) or by cascading the reference signals (see Fig. 9.11).
There is, of course, a significant difference in the way in which

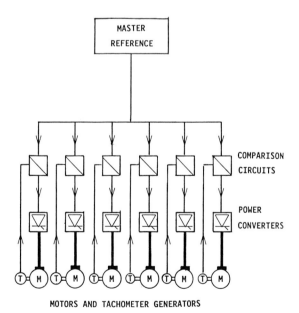

Fig. 9.9 Multi-motor drive system with common reference to all
sections.

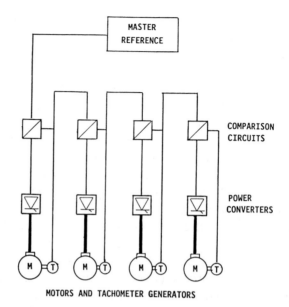

Fig. 9.10 Multi-motor drive system with drives cascaded successively from the preceding section.

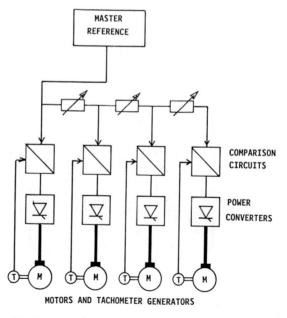

Fig. 9.11 Multi-motor drive system with cascaded references.

the two systems behave. With a cascaded reference each section is effectively isolated from its adjacent section. Where the speed control signal for a section is received directly from a preceding unit the section follows any variation in speed occurring in that unit. The elasticity of the product and the nature of the process largely determine which is more suitable for any application.

To summarize, a cascaded speed control system is more often met with in cases where line speeds are low and the material being processed is relatively elastic or strong. High speed production lines tend to employ very sophisticated controls and cascaded reference systems are generally used in conjunction with high performance drives. The system of cascaded control is often referred to as 'Series speed control'. A combination of both series and independent intersection speed control is sometimes employed. The independent feature is chiefly used for setting up the process or where difficult conditions of manufacture dictate the need for independent adjustment with the system transferring to series control on having established the correct operating criteria.

Chapter 10
Trends

Considerable work continues on the development of variable speed drives and it is interesting to speculate on the way in which designs will change or develop over the next few years. Twenty years ago it was predicted by 'experts' that within 10 years d.c. drives would completely give way to a.c. types and, notwithstanding the lesson of time, a new group of experts are still predicting the demise of the d.c. drive in the next ten years. It is, however, unlikely that d.c. systems will ever be totally eclipsed, providing as they do, a simple robust drive system with many of the features demanded by industrial processes inherent in their design.

It is true, however, that a.c. drives are becoming a more acceptable alternative to d.c. drives as development takes place. Probably the most significant advance in recent years has been the development of specialized circuits for a.c.-controlled drives that have the ability to duplicate many of the desirable features of d.c. drives. The innovative 'torque angle' or 'vector' control system now being extensively used by manufacturers of a.c. drives is a typical example of this. The relatively more complicated circuits required to operate a.c. drives continue to be a major drawback and it is anticipated that more a.c. drive systems will come onto the market with sophisticated fault diagnosis either built into the drive converter/inverter package or available separately with a 'plug in' feature which would permit rapid identification of a fault without the need for specialized knowledge.

Modular design

Modular design is now regularly featured as a sales point and is worth examining in some detail. As originally conceived, this was the process of dividing up the various circuits into physically separable units and connecting them together via easily demount-

Plate 10.1 Modular digital controller used for controlling the output of an inverter or convertor in a speed-controlled drive system. This type of module can be utilized in retrofitting or updating older analogue controllers. (Courtesy of Control Solutions Ltd.)

able plugs and sockets. Rack-mounted circuits were combined by mounting on a common back plane which carried not only the connection from circuit area to circuit area but also the power

Plate 10.2 Operator's control keypad for a digital control system. (Courtesy of Control Solutions Ltd.)

supply to each circuit module. This system permitted the user to diagnose the faulty circuit module and replace it with a spare unit which could be made freely interchangeable without adjustment.

A more recent utilization of the modular concept treats the power converter or inverter with its associated firing and snubber circuits as a complete module. All the drive logic and speed computing circuits are contained in a separate digital computing and control module which supplies the system intelligence and which can also provide high-level communication to a centralized remote area controller. This technique has proved very successful in providing a method of upgrading outdated analogue controlled drives, it is a very cost-effective method because the high-power components are re-utilized.

The development and incorporation of integrated circuits has reduced the number of discrete components being used in the circuit design, and this has led directly to the current tendency among some manufacturers to consider the entire converter or inverter and its associated electronic circuits as a module. Failure of a module in this instance would imply changing the entire unit, thus permitting its return for repair or replacement to be carried out remotely from the operating location. In this case it would be necessary to readjust the drive feedback and control circuits to ensure a return to optimum performance.

This potential disadvantage compared with the modular circuit replacement concept has been recognized by drive designers and future drives will almost certainly incorporate some form of rapid optimization or even automatic self-adjusting optimization circuits. The incorporation of such circuits does mean, however, that specialized knowledge of the function of the various circuits that go to make up the complete package is no longer necessary for setting up, commissioning and fault diagnosis.

Advantages and disadvantages

There are both advantages and disadvantages for the user. One disadvantage is that as this specialized knowledge recedes from users, their ability to make innovative changes to their plant will become progressively more limited; they will become less able to exercise control over the development of the drives and have to rely increasingly on the research and development activities of manufacturers. The contribution made by the user has always been a force for change and it is to be hoped that this input will not be lost.

By careful design and selection of components, manufacturers are also now able to offer extended warranties and better performance guarantees although in the case of most continuous process industries an extended warranty cannot compensate for the loss or potential loss of production if the plant is standing idle while waiting for a replacement to arrive from a supplier. In these industries an extended warranty cannot dispose of the necessity to have extensive on-site spares coverage. In the case of a user with a multiplicity of sizes of drive, total coverage by complete converter/inverter modules could be an expensive proposition when compared with a circuit modularized drive in which sub-components can be extensively duplicated and thus reduce the cost of spares holding without reduction in security of operation.

Digital technology

A recent development that is certain to affect the future of drives is the increasing use of digital technology in both the amplification and control circuits. Digital circuits rely for their performance on arithmetical counting, multiplying and dividing procedure. Until

Plate 10.3 An a.c. drive module featuring flux-vector control to achieve high dynamic performance coupled with an integrated control concept incorporating data logging, local logic and serial communications. (Courtesy of Cegelec Industrial Controls Ltd.)

recently digital technology for dynamic high speed processes was not considered ideal because of the limitations on the ability of available hardware to perform at a sufficiently high speed. Hardware is now available which can perform these functions at very high speeds and this has permitted a more extensive application of this technology for drives. Digital circuits have the advantage of being able to operate without drift and to an extremely high resolution; this enables excellent accuracies, short and long term, to be achieved.

Improved power factor

The increasing use of variable speed drives has also drawn attention to the need for drives with improved power factor. This area of development is likely to be of increasing importance, not least because of the current preoccupation with improved utilization of power. Harmonic generation and other supply disturbances are also areas that are receiving more attention; this is due in no small measure to the increasing use of computers and other supply disturbance sensitive equipment. Drive manufacturers are being asked more frequently to become better acquainted and involved with clients' supply systems and to conduct analyses of (and propose solutions to) possible problems arising from the use of solid-state drives.

Future performance of drives

The following points, while not exhaustive, would surely figure prominently in any speculation on the content of the specification of the drive of the future:

- Power factor approaching unity.
- No voltage spikes in the motor supply waveform.
- Ninety-five per cent efficiency or better.
- No harmonic distortion of the supply.
- No harmonic currents in the motor supply current.
- Modular construction for ease of maintenance.
- 'On board' self-diagnosis of maloperation.
- Low first cost.
- Ability to utilize standard motors.
- Constant torque over entire speed range.
- 100 : 1 speed range as standard.
- Speed control performance better than 0.01%.

Some manufacturers claim that they have already met all the above criteria; the cautious buyer would do well, however, to satisfy himself that all the points have been met without any qualifications which may detract from what appears an ideal solution.

New devices

In considering trends in the design of drives it would be inadvisable to discount the emergence of new converting, rectifying or amplifying devices. Indeed, considerable sums of money are being spent every year in an effort to produce new components of this nature. It is highly possible that a device of a sort to revolutionize the industry will emerge, but as this is a highly competitive field manufacturers tend to keep such information very much to themselves until ready to release new products.

Drive communications

In drive systems much is being done to render the drive more responsive to communication with high speed data highways and other networks. In this area the use of digital technology has assisted greatly. Drives can now incorporate circuits which enable them to be configured as outlying stations of a communications network; this enables information to be transmitted to and from the drive over the network. This permits complex machinery to be controlled, and its performance (as well as that of the drive) assessed, by a remote computer which can also integrate complete manufacturing processes with other plants which may be at different locations. The drive in these systems, as far as the computer is concerned, behaves like a remote information input/output station. Trends in load demand, inactive time, stoppages, etc., can all be time-logged and analysed. Data can be sent to a local operator informing him of a requirement for a recommended maintenance procedure, and the ability of the drive to communicate directly with computers or logic controllers will become commonplace.

The future

Drives of the d.c. type were the primary ones supplied for variable speed application for the better part of 80 years and each new development was hailed as the ultimate in variable speed drives. Development commenced with simple field control of a d.c. motor, proceeded through Ward-Leonard variable voltage d.c. control and thence to power magnetic amplifiers and thyristor-controlled power circuits. At each stage considerable thought was expended on the question, 'Where next?'

One thing is certain: in this rapidly changing technological world things are unlikely to stay the same. Perhaps a new device will emerge to supplant the ubiquitous thyristor and transistor – or even a new design of motor.

Appendix
Conversion data for drive calculations

Length: 1 in. = 25.4 mm
               1 ft = 0.3048 m

Area: 1 sq in. = 6.45 sq cm
               1 sq ft = 0.093 sq m

Volume: 1 cu in. = 16.387 cu cm
               1 cu ft = 0.0283 cu m
               1 gallon (imp) = 4.5459 litres

Acceleration: $1 \, ft/s^2 = 0.3048 \, m/s^2$

Power: 1 hp = 0.746 kW

Pressure: $1 lb/in.^2 = 0.0703 \, kg/cm^2$

Speed: 1 ft/s = 0.3048 m/s

Moment of inertia: $1 \, lb/ft^2 = 0.042 \, kg \, m^2$
                         $1 \, lb \, ft^2 = 0.412 \, Nm^2$ (newton metres)

Energy: 1 ft/lb = 1.367 joules
               1 ft/lb = 0.139 kgf (kilopound metres)
               1 joule = 1 watt s = 1 newton metre

Torque: 1 lb ft = 0.139 kgf/m
               1 lb ft = 1.367 Nm

Web tension: 1 lb/in. = 0.179 kgf/cm

Weight: 1 lb = 0.4536 kg

Glossary

A.C.: Alternating current.

Ambient temperature: The temperature of air, water or other medium in which the equipment is to be operated.

Armature: The rotating part of a d.c. motor.

Base speed: The nominal speed for which a motor is designed and at which the rating is usually defined.

Closed loop: A control system with feedback.

Cogging: Impulsive rotation, usually occurs at low speeds and is associated with discontinuity of the load current.

Converter: A device or circuit for changing a.c. supplies to d.c.

Current limit: The maximum value of electrical current to which a system is controlled usually for protection of the converter or the device to which the converter is connected.

Damping: The process of reducing oscillation or overshoot of a closed loop control system following a system disturbance.

D.C.: Direct current.

Dead band: A term used to describe the period between a converter operating in the forward direction and changing to the reverse direction. This term is also applied to the amount of change in input to a system that can occur before a change in output takes place.

Deviation: The amount by which the controlled parameter diverges from the desired value.

Diode: A semiconductor device that can conduct electricity in one direction only.

Drift: An uncontrolled deviation from the preset value usually associated with a temperature or other external change.

Duty cycle: The operating cycle of a process which defines the periods of inactivity and operation.

Encoder: A device that produces a series of pulses which are directly related to rotational or positional movement and providing a feedback of these values for a control system.

Error: The deviation of a control system from a desired or preset value.

Feedback: The signal fed to a control system from the controlled element which actually defines the state (speed, position) of that element.

Field: The part of a d.c. motor that produces a magnetic flux against which the flux created by the load reacts – usually the fixed part of a motor.

Filter: A circuit or device that has the property of accepting or rejecting a specific frequency.

Form factor: A factor used to define the deviation of an electrical wave form from a pure sine-wave. In mathematical terms it is the ratio of rms (root-mean-squared) value to the average value of the waveform and defines the relationship between the heating effect of the waveform and its ability to produce useful torque in a motor.

Gate: The element in a controlled rectifier that is used to control the conduction in a forward direction.

GTO: Gate turn-off thyristor is a semiconductor device that can be turned on or off via the gate circuit.

Head: The pressure that exists due to the height through which a fluid is raised. Usually expressed in terms of a height of water, e.g. 10 metres w.g. (water gauge).

Heat exchanger: A device or piece of equipment with which heat can be removed indirectly from an enclosure that is not open to the atmosphere.

Hysteresis: The effect exhibited in a material which tends to retain its state and which energy is required to change. Hysteresis loss is a measure of this energy.

Induction motor: An a.c. motor in which the rotational impetus is derived by induced currents between the fixed and rotating elements.

Inertia: The reluctance of a body to change its state of rest or constant motion.

Instability: The state of a control system when the output and input do not correspond, i.e. the output of the system is not acting in conformity with the command signal.

Interpoles: Auxiliary magnetic poles added to a d.c. motor to assist commutation.

Inverter: A device or circuit for changing a.c. currents to d.c. or variable frequency a.c.

IR compensation: A compensating signal designed to take into account the variable voltage drop in a d.c. motor due to a change in armature current. This factor assists in compensating for the effect of load changes, i.e. the regulation of the motor.

Jog: The process of moving a motor by small momentary increments.

Kinetic energy: The energy possessed by a body in motion.

Negative feedback: A system where the feedback signal opposes the reference signal.

Open loop: A system without feedback.

Overload capacity: The ability of a motor or drive system to withstand loads in excess of its normal continuous rating. Usually specified as a percentage of its nominal rating; sometimes related to time.

Overshoot: The amount by which a controlled level is exceeded after a change to the input to the system.

Power factor: The ratio of actual power (watts) to apparent power (volts × amps).

PWM (pulse width modulation): A method of constructing an artificial sine-wave from a series of pulses.

Reactance: A property possessed by an electrical circuit that opposes a sudden change in the current flowing in that circuit.

Regulation: The change in speed or other controlled variable following a change in another parameter, e.g. change in speed for a change in load.

SCR (silicon controlled rectifier): A solid state rectifier with the ability to conduct when a signal is applied to one of its electrodes.

Service factor: A factor applied to the rating of a motor to indicate its ability to cope with loads different from the nominal rating.

Slip: The difference between synchronous speed and actual speed of an induction motor.

Stability: A measure of a system's ability to remain at a preset condition of operation.

Switched reluctance drive: A drive system which depends for its operation on the residual magnetic properties of iron or steel and a synchronized switched excitation circuit.

Torque: A turning force which, when applied to a shaft, tends to cause rotation.

Transient: A momentary deviation from a steady state condition.

Vector: A method of representing magnitude and direction of a force or other quantity.

Index

accelerative response, 122
a.c. motors, 14
  cage, 15
  induction, 15
  Kramer, 60
  N−S, 57
  pole-changing, 21, 56
  series, 16
  slip-ring, 16
  stator/rotor fed, 18
  synchronous, 17
armature reaction, 8, 12, 69
armature resistance control, 71
applications, drive suitability for, 114

braking
  regenerative, 80, 81
  resistance, 56, 71, 81
break-away torque, 115
brushes
  grade, 13
  position, 12
  selection, 13

cage motors, 15
calculation of power requirements, 111
cascaded control, 123, 125
characteristics
  of control systems, 31
  of load, 101
choppers, 71
closed loop control, 25, 27, 30, 33, 36
coilers and uncoilers, 108
communication with drives, 132
commutation and commutators, 10
  dead band for commutation, 12
  effect of vibration on, 14
  function of, 11
  glazing, 13
  pitting on commutators, 13
  problems with, 13
  sparking, 13

comparisons of drive types
  a.c., 97
  d.c., 98
compound wound d.c. motors, 10

control system response, 33, 117
converters
  cyclo-, 65
  multi-phase, 98
  single-phase, 51
conveyor drives, 106
cooling
  drive enclosures, 93
  motors, 53
  semiconductors, 49
critical damping, 32
current fed inverter, 63

dead band commutation, 12
digital control, 129
diodes, 38
direct current motors, 5
  brush position, 12
  commutators and commutation, 10
  compound type, 10
    separately excited type, 9
  series type, 5
  shunt type, 7
distortion of electrical supplies, 87

efficiency, 97, 99, 120
electromagnetic compatibility (EMC), 89
emission and radiation, 91
EU directives, 90
  compliance with, 90
  technical construction file, 94
  testing, 95
examples of control systems, 28
extruders, 107

fan drives, 105
feedback, 26
feed forward control, 25
field control of speed, 69
forces acting on a conveyor, 107
frequency
  relationship to speed, 21
  speed control, 61
fuses
  use in protecting semiconductors, 45
  construction, 47
  characteristics, 47

gate turn-off devices (GTO), 44
glazing of commutators, 13
grading of brushes, 13

harmonics, 83
  analysis, 84
  magnitude, 85, 86
  reduction, 87–8

induction motors, 15
  characteristics, 55
  starting, 55
  torque, 54
  variable speed applications, 53
inertia, 75, 77, 115
inverters, 49
  current fed, 63
  voltage fed, 62
  quasi square-wave, 64

Kramer speed control, 60

motors
  a.c., 14
  d.c., 5
  reluctance, 22
modular design, 126
  advantages and disadvantages, 129

N–S variable speed motors, 20, 57

open loop control, 25, 27
overdamped, 32

peak inverse voltage (PIV), 40
performance specification, 34
  steady state, 34
  transient, 35
pitting of commutators, 13
pole changing motors, 20, 56
poles, speed relationship in a.c. motors,
  20
position of brushes, 12
power factor, 131
power/flow/pressure characteristics
  fans 105–6
  pumps, 104
protection of semiconductors, 45
pulse width modulation, 65
pump drives, 102

quasi square-wave inverter, 64

rectifiers, 38, 39
regenerative braking, 80
reluctance motor, 23
resistance speed control
  a.c. motors, 56
  d.c. motors, 71

resonance, 116
response of control systems, 117
rotary converters, 99
running cost, 119

sectional drives, 122
selection of brush grades, 13
semiconductors, 37
  protection, 45
separately excited motors, 9
series a.c. motors, 16
series d.c. motors, 5
shunt motors, 7
single-phase converters, 76
slip, 16, 55
slip-ring motors, 16
  speed control, 56
slip recovery speed control, 59
snubber network, 47
soft starting, 55
sparking of commutators, 13
starting torque
  consideration in drive selection, 103,
    115
  d.c. motors, 6, 8
  induction motors, 55
  synchronous motors, 18
static head, effect on pump performance,
  105
static Kramer system, 60
stator/rotor fed motors, 18
steady state performance, 34
stresses due to resonance, 122
synchronous motors, 17

transient performance, 35
transistors, 40
thyristors, 42
torque
  induction motor, 54
  d.c. motors, 6, 8
three-phase converters, 77, 78
torque angle control, 126
torsional rigidity of load, 116

underdamped control, 32

variable frequency control, 61
vector control, 126
vibration, 14
voltage-fed inverter, 62

Ward Leonard-Ilgner system, 71, 75
wave form
  single-phase converters
    full-wave, 78
    half-wave, 78
  three-phase converters, 79